Environment Science: Research and Analysis

Environment Science: Research and Analysis

S.R. Manjunatha

2008

CYBER TECH PUBLICATIONS
4264/3, Ansari Road, Daryaganj, New Delhi-110002 (India)
Ph.: 011-23244078. Fax: 011-23280028
E-Mail: cyberpublicationsdelhi@yahoo.com

ENVIRONMENT SCIENCE : RESEARCH AND ANALYSIS

S.R. Manjunatha

First Edition 2008

Published by :

G.S. Rawat for **Cyber Tech Publications**
4264/3, Ansari Road, Daryaganj, New Delhi-110002 (India)
Ph.: 011-23244078. Fax: 011-23280028
E-Mail: cyberpublicationsdelhi@yahoo.com

Printed at : **Amit Enterprises**, Maujpur, Delhi, (9810878184)

Preface

Environment is anything that surrounds us from all sides and has a great bearing on our everyday life. Human being is the product of his environment and is deeply influenced by it. But at the same time he has left an indelible impression on environment by his activities. He has been exploiting the natural resources to the hilt not only to meet his needs but also for his greed. Equipped with technological advancement with each passing day, he has been trying to make his life more comfortable and raise his standard of living by accumulating luxuries around him.

Environment is not bound by political boundaries drawn by man. Rather it is a global phenomenon and needs attention at the global level. But now, man has become aware of the environmental crisis committed by it.

India is also no exception to this. It is a developing country where rapid population growth and wide-spread poverty are two main causes of environmental degradation. In this book all the related topics are briefly discussed in a lucid language and easily understandable manner. I hope this book will serve its purpose.

"SAVE ENVIRONMENT SAVE LIFE"

—*Author*

Contents

1

Introduction of Environment Science

INTRODUCTION

Throughout the entire Universe, and throughout the entire history of the Universe, the Earth is the only planetary body with the appropriate conditions for life as we know it to be developed and sustained. For that reason, Earth is a very special planet. Earth is also a very fragile planet, and if too severely abused, life that has evolved and sustained for over three billion years will terminate.

For most of these three billion years, life forms have lived in narrowly-defined ecological niches, with limited geographic mobility. As a result, they lived in harmony with their environment and had little impact on the planet as a whole. However, in the last three million years, Homo sapiens have evolved. Unlike most "lower" life forms, we alter uninhabitable or marginally habitable environments to better suit our needs. Consequently, humans have made the entire planet their ecological habitat. In our efforts to maximize space, however, we have too often made decisions that have irreparably altered many of our environments.

Many earth scientists are aware, and are becoming increasingly more so, that we are living at perhaps the most critical period in Earth's history. The impact of the human race is so significant that if we do not change our ways, Earth will suffer irreparable damage. Human population is increasing at an alarming rate; many of our non-renewable, life-sustaining and other natural resources will be depleted within two generations. Air, water and land pollution is increasing, and decisions on land use are being made strictly on economic and political grounds, when they should have input from professional earth scientists. Unfortunately, most people who are not professional earth scientists are unaware of the severity of the issues we face. Because they are not aware, they are unconcerned and unlikely to take action.

Here we will discuss the following earth issues:

- The role and impact of human population growth
- Goals for environmental education
- When and for whom such a course would be appropriate

ROLE AND IMPACT OF HUMAN POPULATION GROWTH

When asked, most professional earth scientists would probably identify the following "mega-issues" as our most serious environmental issues:

- Human population growth
- Consumption and depletion of life-sustaining and other non-renewable natural resources
- Pollution
- Land use

While it could be argued that any one of these is more important than the others, the fact is that these four "mega-issues" are so intimately interconnected that it is impossible to address one effectively without addressing some or all of the others. For example, in the 1980's, South Carolina experienced a rebirth of its gold-mining industry. However, in spite of the promise of new jobs and an increase in the tax base, many people opposed the opening or reopening of the state's gold mines. Why was this issue so controvertial? The mines could re/open because demands of an increasing population brought about a greater consumption of this non-renewable resource. The relatively low-grade ore in South Carolina suddenly became economical to extract. However, a great deal of rock material would have to be processed, leaving a great amount of waste, or spoil material. People opposed the opening or reopening for the following reasons:

- Potential for air, water, and/or soil pollution
- Land degradation and environmental restoration (land use)

While it can be shown that effectiveness in addressing any of the four issues is contingent upon addressing the other issues as well, it is very difficult to find an issue for which human population growth is not a major factor. For that reason, throughout this course, we will return repeatedly to the issue of human population growth.

GOALS OF ENVIRONMENTAL EDUCATION

Most people who are not professional earth-or life-scientists may not be aware of the seriousness of the issues we face. Too often, we as educators feel that our responsibility ends at developing awareness. Unfortunately, raising the awareness of others may not be enough.

These issues are sufficiently critical that it is imperative that people begin to be proactive in addressing them, (i.e., taking some type of personal action, like writing to Congress). After becoming aware that an issue exists, a person must become concerned, and develop a sense of urgency about an issue before he or she is prepared to take any action. It is not realistic to hope that everyone will become aware and concerned about all issues, but it would be very unfortunate if someone left this class without developing that sense of urgency and a desire to do something about at least one issue.

Many of these issues are so complex and so global in nature that there is a danger that some might develop a sense of apathy or despair, that "there is nothing that I as an individual can do to effect change." We must show that while many of these issues are difficult, complex and even global in scope, actions taken by individuals can significantly aid in their resolution. Therefore, we must leave with a sense of urgent need to take action on these issues and a sense of empowerment that we can make a difference. When that is accomplished, we will have moved beyond awareness to advocacy.

Clearly, if individuals are going to increase their level of commitment to environmental issues by taking personal action, their environmental consciousness must be raised. We recognize five stages of environmental consciousness:

- Holistic-one who seeks harmony with planet Earth, accepts humans as a transitory life-form and recognizes that planet Earth is more important than individual humans or even whole nations or cultures
- Cross-Cultural-one who seeks harmony with others; transcends individual and cultural desires and sees the need for "community"
- Internal-one who seeks harmony with self and is willing to challenge rules of others if these rules are felt to be inappropriate
- Institutional-one who is dependent upon the rules of others; and is intolerant of those who do not follow these rules
- Asocial-one who is strictly "me" centered; does not care about rules unless the rules are a definite benefit to him or her.

We believe that both individuals and societies progress through these stages (from bottom to top), but acknowledge that not every individual and certainly not every society reaches the Holistic Stage. In the United States, for example, we estimate that there are varying numbers of people in the each stage and that the distribution most likely follows a Gaussian distribution, with most people currently in the Institutional or Internal Stage. We as earth science students and educators must strive to reach the Holistic Stage, and we should do everything within our power to assist others to reach the highest possible level. This expands our responsibilities significantly. We must move from being transmitters of knowledge to advocates, and that means we must practice advocacy, both in our classrooms and outside.

It is generally recognized that one of the major aspects of the "education crisis" in the United States is that the public has insufficient knowledge of basic science concepts on which to base intelligent decisions on critical social issues, including critical environmental problems that must be addressed within our lifetime or the lifetime of our children. In order to deal effectively with these issues as a nation, all segments of the general public must strive for a level of scientific literacy never before achieved in history. Clearly, environmental education should begin as early as possible, and-be an important component of life-long learning.

RESOURCES SUSTAINING LIFE

Air

The invisible substance called air touches everything on the earth's surface which is not under water. Humans are in contact with air for every second of their lives. It is an integral part of almost everything we do from growing food to flying an airplane. We can feel air as it moves past us, we can hear it whistle around a building on a windy day, but clean air has no color or taste or odor. Air can be compressed into very small spaces and expand to fill large ones. All of the air in a large house could be compressed into a small tank, and once the house is empty, a cubic foot of air at normal pressure could expand to reach every corner of the same house. Air is composed of tiny molecules of gas, as many as 300 billion per cubic inch of ordinary air. At the most basic level, most living things require oxygen to survive and many of them obtain that oxygen from the air. All living things must dispose of wastes, and many of them dispose of waste gas into the air.

Origin

When the earth was forming over 5 billion years ago, it was initially too hot to have any significant atmosphere. As the earth cooled, it began to create an atmosphere for itself. Volcanic activity, past and present, produces a large amount of gas consisting predominantly of nitrogen, carbon dioxide and steam (water). During the earth's early days, water collected in the atmosphere. As the earth continued to cool, the water started to condense, fall as rain, and fill the oceans. Eventually these oceans could support plant life. The plants used the water and the carbon dioxide to achieve photosynthesis, and this in turn produced oxygen as a by-product. The oxygen began to accumulate in the atmosphere. The activities of plants and animals, as well as other processes, continued to remove carbon dioxide from the new atmosphere, and flourishing plant life continued to add oxygen. Most of the non-reactive nitrogen remained untouched. In this way the current atmosphere evolved to form a thin layer surrounding the earth. The majority of the atmosphere lies within 60 miles (100 kms.) of the earth's surface, but small traces of gas may be found for hundreds of miles above that.

Composition

Constituent	Percent by Volume
Nitrogen (N_2)	78.084
Oxygen (O_2)	20.946
Argon (Ar)	0.934
Carbon Dioxide (CO_2)	0.032
Neon (Ne)	0.001
Helium (He)	< 0.001
Krypton (Kr)	< 0.001
Xenon (Xe)	< 0.001
Hydrogen (H)	< 0.001
Methane (CH_4)	< 0.001

The atmosphere's composition varies slightly at high altitudes, but the vast majority of the earth's atmosphere is a remarkably uniform mixture of 78% nitrogen, 20% oxygen, .9% argon and .03% carbon dioxide. The remaining fraction of a percent is composed of traces of neon, hydrogen, helium and methane. These are the components of "dry" air. The only significant variation occurs in the air's content of water vapor which can range from almost nothing up to a high of 4%. At 4%, the air is saturated with water vapor. This is often expressed as a "relative humidity" reading of 100%, meaning that the air is holding 100% of the water vapor it is capable of holding at that time. Other minor variables are ozone (O_3), sulfur dioxide (SO_2), carbon monoxide (CO) and water (solid and liquid). Of these variables, ozone is probably to most important.

Soil

Soil has been described as the bridge between the living and non-living portions of the earth. The Earth's crust is made of rock. If its outer portion were not changed in some way, the entire surface of the earth would be bare rock. Only a very few living things could survive under these conditions. Most plants cannot grow without some sort of soil, and no animals can survive without plants. Humans are no exception. We are dependent on soil for the production of food, and thus we are dependent on all of the processes that create and maintain soil. Soil forms as rain, wind, ice and a variety of other agents break down the earth's rocks. Collectively, these processes are known as weathering. Soil is composed of portions of the original rock that are combined with several other components that form during the weathering process. Many of these components are essential for the growth of plants. Generally soil forms in layers known as horizons. The horizon containing the most immature soil is at the bottom close the rocks from which the soil is forming, and the horizon containing the most mature soil is at the top. The type of soil that forms in a particular area, and the thickness of the soil is primarily determined by the climate in which the soil is forming.

Soil has been described as the bridge between the living and non-living portions of the earth. The Earth's crust is made of rock. If its outer portion were not changed in some way, the entire surface of the earth would be bare rock. Only a very few living things could survive under these conditions. Most plants cannot grow without some sort of soil, and no animals can survive without plants.

Humans are no exception. We are dependent on soil for the production of food, and thus we are dependent on all of the processes that create and maintain soil. Soil forms as rain, wind, ice and a variety of other agents break down the earth's rocks. Collectively, these processes are known as weathering. Soil is composed of portions of the original rock that are combined with several other components that form during the weathering process. Many of these components are essential for the growth of plants. Generally soil forms in layers known as horizons. The horizon containing the most immature soil is at the bottom close the rocks from which the soil is forming, and the horizon containing the most mature soil is at the top. The type of soil that forms in a particular area, and the thickness of the soil is primarily determined by the climate in which the soil is forming. Weathering of Rock

Rocks form at very high temperatures and pressures deep within the Earth. When rocks are exposed to the conditions that exist at the Earth's surface (low temperatures, low pressures, abundant air, and abundant water), they react to the new conditions and begin to change. That process of change is known as "weathering" and it is accomplished through two primary sets of processes: "Physical or Mechanical Weathering"- is the breakdown of large rocks into smaller pieces that have the same chemical and mineralogical composition as the parent rocks. This is accomplished by processes such as:

Frost-as water freezes in cracks in the rock it expands and wedges the cracks open further; Growth of plants-the roots of plants find their way into cracks in the rock, expand as they grow and break the rock; and Release of pressure-as erosion wears away the rocks that lie above, pressure is released from the rocks below and they expand very slightly, cracking their outer layers.

The processes of physical and mechanical weathering create openings in masses of rock that act as channels for air and water to reach the rock's interior. At the same time, the fact that rock is broken into smaller pieces means that the amount of 'surface' being exposed to air and water is increasing. Both of these actions facilitate the other primary type of weathering known as "chemical weathering." Chemical Weathering is the chemical alteration of rock and its components to form new materials. The primary chemical processes are:

Oxidation-combination with oxygen to form oxides such as iron oxide (Fe_2O_3);

Solution-becoming dissolved in water. Some minerals dissolve easily in pure water (such as halite-NaCl). Others become more soluble in acidic water. Rain water becomes naturally acidic when it combines with carbon dioxide to form carbonic acid ($H_2O + CO_2 = H_2CO_3$).

Hydrolysis-a chemical reaction between a mineral and the components of the water molecule to create a new mineral. A prime example of this process is the transformation of the mineral 'orthoclase' ($KALSi_3O_8$) into clay [$AL_2Si_2O_5(OH)_4$]

Composition of Soil

About half of the material that makes up a good fertile soil is weathered rock constituents such as sand, clay, silt and small fragments of the original rock. The larger particles such as sand grains and rock fragments are important because they hold the finer particles apart making it possible for air and water to circulate through the soil. The finer particles such as silt and clay are important because of their ability to hold water, and in some cases, to hold essential nutrients for plants. The other half of a fertile soil is "humus", the dark-colored material that is derived from the decay of organic matter from pants and animals. The presence of humus helps a soil to retain moisture, but its greater importance is as a source of essential nutrients.

Nutrients, often referred to as "trace elements", are minute quantities of common elements such as iron, sulfur and nitrogen without which plants cannot grow or survive. Some nutrients are a product of the weathering of rock, others are released by the decay or organic material. For example, "proteins" are an essential part of all plants and animals. Proteins are composed primarily of carbon, hydrogen and oxygen but they also contain small quantities of nitrogen, sulfur, and often iron, phosphorous, or iodine. The carbon, hydrogen and oxygen necessary to make the proteins are readily available in air and water, but the other components are more rare.

Nitrogen is abundant in the atmosphere, but it cannot be used by plants until it has been combined with another element such as oxygen (to form nitrates) or hydrogen (to form ammonia). Once the combination has occurred the nitrogen is said to be "fixed" and it can be taken into a plant to become part of a protein. Fixed nitrogen is an important nutrient. When the plant dies and is broken down by bacteria, the decaying organic material releases the fixed nitrogen to be used again in the next generation of plants. Other nutrient follow a similar cycle.

Many originate in rock and become available as the rock weathers. Released into newly forming soil, small quantities of iron or phosphorous may become important nutrients for plants. Once a part of the pants, the nutrients are retained until they are recycled by the processes of decay. After a plant has decayed, its nutrients are temporarily stored in the soil. Some important nutrients, such as nitrates and ammonia, are actually produced by bacteria that live in the soil. For this reason erosion must be seen as a very serious problem. When soil is washed or blown away, the area has not only lost the soil, but it has lost a valuable supply of nutrients as well.

Soil Horizons

A hole dug into deep soil will usually reveal a number of distinct layers. These layers are known as "Horizons" and represent the various stages in the development of soil. Since soil forms from the weathering of rock, it forms from the bottom up. The newest and least mature soil is at the bottom adjacent to the rock from which it is forming. This soil is composed of sand, clay, and substantial chunks of the weathering rock. There is very little organic matter in this layer. This bottom layer soil is known as the 'C' horizon. The 'B' horizon lies in the center and represents extensively weathered material containing some sand, but usually a very high percentage of clay. The 'B' horizon does not contain much organic material, but it does contain some of the components that have been washed (leached) out the soil above. On the top is the 'A' horizon or "top soil". This is the active area of the soil and the only one that contains a significant amount of organic material. Some the organic material is live plants, the remainder is humus.

Food Production

Healthy soil is essential for human food production. Currently the Earth has approximately 3.7 billion acres of land with sufficient water and soil to grow crops. This represents 11% of Earth's total surface. Hypothetically, another 3.7 billion acres could become usable if improvements were made to the land. This total amount of land, properly managed, should be adequate to feed the people of the world. Tragically, however, tens of thousands of people die each day as a result of starvation. Examining the food production and consumption habits of the world can give us some insight into the environmental problems associated with food production and hunger, and the sad irony of starvation in the midst of plenty.

In an attempt to meet human nutritional requirements, the world's major crops consist chiefly of wheat, rice, corn, and potatoes. In many developing countries, barley, sweet potatoes, and soybeans are also important crops. As a protein source, domesticated animals constitute another vital component of the human diet. The animals most commonly raised for consumption are cattle, pigs, sheep, horses, poultry, mules/ asses, goats, camels, and buffaloes. Grazing these animals uses 7.4 billion acres of land.

Nutritional Requirements

These different foods form the foundation of a healthy diet. On average, people require 2400 calories per day. These calories are derived from carbohydrates, which provide energy; proteins, which form muscles and organs and are necessary for growth and development; and fats, which build cells and act as insulation. Iron and vitamins are also necessary for the proper functioning of the body. The majority of calories consumed in a day should be obtained from carbohydrates (grains), with a much smaller percentage coming from fats and proteins. The extent to which these guidelines are followed has implications for both an individual's health as well as the world's hunger problems.

Trophic Levels

"Ecology" is the study of the structure, function and behavior of the natural systems of the biosphere. "Physics" is the study of matter, energy, motion and force. An understanding of some basic ecology and physics is necessary in order to see the connections between our food choices, world hunger and the environment. Humans, like all other organisms, require energy to live, grow and reproduce. The ultimate energy source for the Earth is the Sun. There are several organisms that draw energy directly from the Earth, but their numbers are inconsequentially small. The energy that fuels the vast majority of the biosphere (the living systems on Earth) comes from the Sun and is converted by plants into useful forms-for themselves and all other organisms.

Most of the plants that collect and store this energy are located in specific ecosystems on Earth. Some of the most productive areas are coastal zones, estuaries and rain forests. Unfortunately, these same areas are currently under environmental pressures due to human encroachment. Certainly, any decisions made regarding the preservation and management of these ecosystems will have lasting effects on the sustainability of Earth.

Through the process of photosynthesis, plants use the Sun's energy to produce sugars, which then combine to create carbohydrates. Plants produce these products to meet their own needs, and then store them in seeds, roots, and sap. When plants are eaten by an animal, these materials (carbohydrates) are used by the animal to maintain its life processes. If that animal is eaten by another, the energy is passed on again. This is known as a "food chain", composed of producers (the plants) and consumers (the animals). The position a consumer occupies on the food chain is known as its "trophic level". The first animals to eat the plants are called primary consumers. The animals that eat primary consumers are secondary consumers and are described as being on the next trophic level. In some cases this extends up to third and fourth level consumers. For example, a gazelle that eats grass would be a primary consumer. A lion that eats the gazelle would be a secondary consumer, and if a man killed and ate the lion, he would be a tertiary (third) consumer. The man with a belly full of lion would be at the third trophic level.

The quantity of energy created by producers is never fully useable by its subsequent consumers. The transfer of energy process is quite inefficient; in fact, an average of only 10% of the energy is passed on in a usable form during each transfer to a higher trophic level. Most of the energy is lost as heat, just as our car engine produces so much heat. Thus, because of the decreasing amounts of energy available at each level there are fewer animals at the higher trophic levels.

Human diets can be either energy "efficient" or "inefficient" depending on the foods eaten. Diets rich in grains and supplemented with minor amounts of chicken or pork are "efficient", because less energy is "wasted" during transfers up through the trophic levels. Diets high in animal content (especially grain-fed beef) are highly "inefficient", because so much energy is lost. A closely connected problem is that feeding animals meant themselves to be eaten ties up huge amounts of farm land. When crop land is used to produce grain to feed cattle, approximately 90% of the energy originally present in the corn is lost by the time a human body "burns" the calories in the beef. Many more people could be fed-and on a healthier diet-if the corn went straight into the mouths of people rather than into the mouths of cows. It is logical that one way to combat hunger and increase sustainability is to encourage consumption at lower trophic levels. Another way to deal with the shortage of food is to grow more food per acre of available land. Attempts to do this are known as green revolutions.

TRANSFER OF ENERGY

Some of what we know about the transfer of energy from one form to another is described by the First and Second Laws of Thermodynamics.

The First Law of Thermodynamics states: "During a chemical or physical change energy is neither created nor destroyed; however it may change form and may be moved from place to place". For example, plants transform radiant energy from the sun into chemical energy which is stored within the plant. When eaten by an animal that chemical energy may be transformed into heat.

The Second Law of Thermodynamics states: "With each change in form some energy is degraded to a less useful form and given off to the surroundings as low-quality energy (heat)". This means that transformation of energy is not very efficient, and that much of the energy is often lost in a less useful form. For example, an automobile extracts chemical energy from the fuel it burns in order to get the energy to move along the highway. A great deal of that energy is lost as heat, however, making the engine hot enough to require a "cooling system".

A light bulb is another example; the bulb is converting electrical energy into light, but loses so much energy as heat that it is usually too hot to touch.

GREEN REVOLUTION

Throughout human history people have attempted to defeat hunger by increasing the productivity, or the "yield" of the land. Historically, farming has been a labor-intensive endeavor that depended mainly on the natural conditions of the soil and the availability of water. Modern farming has become highly mechanized, and has attempted to overcome disadvantageous natural conditions with a variety of techniques including chemical pesticides, petroleum-based fertilizers, and modern irrigation methods. This transition to modern farming, sometimes referred to as a "green revolution", coincided with the development of strains of crops which produce much higher yields. These "high-yield" varieties of crops have allowed many countries to greatly increase food production and to feed many more of their people. On the other hand, these crops do come at a high cost: first, the seed is typically more expensive than native varieties; and second, such crops often require extensive fertilization and irrigation, which may be beyond the means of the average farmer. This has encouraged a trend toward the consolidation of farm land into plantations belonging to wealthy landowners or large companies. The production of high-yield crops has also often displaced diverse native crops which had specifically adapted to their particular locales. Native crops are inherently less susceptible to local pests and diseases, and are often more nutritious than their hybrid counterparts. The bottom line, unfortunately, is that in 1998, despite these and other efforts to produce more food and feed the world's people more efficiently, a large percentage of the earth's human population is underfed..

HUNGER AND STARVATION

Each year forty million people die as a result of having insufficient food. Some deaths are the result of diseases which afflict the malnourished, while others are due to plain and simple starvation. Malnourishment comes at a high price, both to those afflicted and to their community. The malnourished have a decreased resistance to disease, which results in higher death rates, and to shortened life expectancies. On a daily level, malnutrition causes fatigue in individuals, which leads to decreased ability to learn or to perform productive work. This in turn contributes to more poverty which leads to more hunger, and the cycle continues.

The cause of hunger is not, ironically, inadequate food or a shortage of arable land on which to grow it. The world has enough land, if properly managed to adequately feed the world's current population. It is estimated that ten billion people (twice the current population) could be fed if all the corn and root crops currently fed to animals were used for human consumption. For humans to eat animals at or near the top of the food chain is an inefficient use of the planet's resources. Hunger is the direct result of social and economic causes. Hunger exists because of inequitable and inadequate distribution of food. In many developing countries, the best farm land is in the hands of a few wealthy landowners. Poorer people are at the mercy of these landowners to get access to the prime land at a high cost; otherwise they must try to scratch out a living on marginal land. The landowners discourage the latter because they then lose their cheap labor, and cheap labor is necessary to produce cash crops. Cash crops are generally luxury items (such as coffee, tobacco and sugarcane) which are exported for the money rather than to feed the workers. Ideally, the cash crop brings in enough money to enable people to buy food and the equipment to produce more food. However, the profits are often realized in distant countries, and by only a few people. The poor workers where the crops are grown are left hungry and powerless to do much about it.

Theoretically, the people of the world could be physically sustained by the Earth--yet they are not. Hunger, a painful, often lethal physical reality to those experiencing it, is nonetheless actually a social problem, and as such is part of a complex web of social, political and environmental issues. As the population continues to expand exponentially, the growing masses struggle to feed themselves. If good land is reserved for cash crops, the underprivileged are forced onto land unsuitable for agriculture. They fight to use this land until it becomes totally unusable and then move on to another location and repeat the process. These practices result in deforestation, overgrazing, soil erosion, desertification and drought: all problems which are difficult-at best-to reverse. Effective soil management is one of our few viable solutions.

WATER RESOURCES

Water is one of the basic resources that sustain life on earth. Without water, no plants or animals could survive. Some organisms extract water from the air, others get water from the oceans. Humans can do both of these, but like many plants and animals, we receive the majority of our water from land. Water on land is found as surface water, water in lakes, streams and swamps; and ground water, water found within the rocks and soil below the surface. The majority (97%) of the earth's water is found in the ocean. Since ocean water is salty, it is not readily available to those plants and animals that live on land and thus have not developed an ability to use salt water. The earth's next largest stock of water, about 2% of the total supply, is frozen as glacial ice in such places as Antarctica and Greenland. This is fresh water, but it is obviously difficult to use unless it is melted. Finally, about 1% of the earth's water is found on land as either ground water or surface water. This figure includes all of the water on land including lakes and streams, and all water below the surface even including moisture in the soil. Realistically, humans have ready access to only about 1/2% of all of the earth's water. This figure may seem distressing, and it should remind people that water is a limited resource, but it should not be seen as the last word on the subject. To fully understand the earth's resources of water, one must consider how water moves through the hydrologic cycle, where the available resources are located on and in the earth, and how the current human demand compares with the available supply.

HYDROLOGIC CYCLE

The Hydrologic Cycle is composed of three or four phases: Evaporation, Condensation, Precipitation, and perhaps, Runoff.

The heat from the sun causes water to evaporate from the ocean and from plants and land surfaces. As the vapor rises in the atmosphere, it encounters colder temperatures and begins to condense forming small droplets of water. When the droplets are heavy enough, they begin to fall as precipitation back to the earth's surface.

If the precipitation falls in the ocean, the hydrologic cycle is complete, but the water has not become available to plants and animals on land. If the clouds are over land when the precipitation falls, the water begins to flow back into the ocean as runoff and is temporarily accessible to land plants and animals. The hydrologic cycle is an important means by which water is distributed, and it is also an import way by which water is recycled. When water evaporates, most of the materials dissolved in the water are left behind. In this way, dirty water is cleaned and made available for reuse. The rate at which natural and human process can clean water determine whether water can be considered a renewable or a non-renewable resource.

SURFACE WATER

Surface water is precipitation that does not soak into the ground, or return to the atmosphere by evaporation or transpiration. Geologists call this water "runoff" since it is in the process of running off the land surface and ultimately back into the ocean. As it progresses toward the ocean, the water it is temporarily stored in lakes and streams. About half of the water humans use comes from surface water. It is readily available, and has been used by humans in a great variety of ways since our appearance on earth.

Unfortunately, surface water has also provided a readily available site for dispose of waste and humans have used it that way throughout history. The supply of surface water available for use by humans as well as by plants and animals is the result of a delicate balance between "input" from precipitation and "output" through use, pollution, and loss to other reservoirs of water (the ground, the ocean, etc.). The Colorado River provides an example of what can occur if this balance is not maintained.

Colorado River

The Colorado River flows 1,470 miles through Colorado, Utah, Nevada, Arizona and Mexico until it empties into the northern tip of the Gulf of California. Before it was altered by human construction, the Colorado carried about 10 million acres feet of water (enough water to cover 10 million acres to a depth of one foot) each year past the point where it crossed the border into Mexico. The river has always been an important source of water in the arid areas on the Mexican Northwest and the American Southwest. As agriculture developed in these areas, and as the population began to increase, the need for an abundant and controlled source of water became more important. Many dams were built along the river to control flooding, to generate hydro-electric power, and to redirect water for use in irrigation. Today the river's flow has been cut from 10 million acre feet at the Mexican border to only 1.5 million as water was siphoned off by projects such as the giant Hover Dam and the All American Canal that irrigate American crops and supply American cities. One point of view is to say that the Americans have "tamed the great river and put it to work". Another view is to say that "we have ruined a great river by overuse", because irrigation projects can diminish a river in several ways. The result is that the flow of the Colorado has not only been cut by 85%, but the quality of this meager flow has been significantly reduced by increases in salinity and pollution.

Irrigation

Approximately half of the world's farmers use some process of irrigation to bring additional water to their crops. Irrigation is required in areas that receive less than 10 inches of rain each year, and additional water can greatly increase production on farms in many other climates. Water for irrigation can be pumped from the ground or removed from surface lakes and streams, but in all cases it has three primary effects on local sources of surface water:

1. *Depletion:* The vast majority of water removed for irrigation is never returned to its source, so irrigation tends to deplete the source from which the water is extracted

2. *Salinity:* All water contains a certain amount of dissolved material. When water is pumped into a field for irrigation, a very large percentage of the water evaporates leaving behind whatever "salt" was dissolved in it. If this salt is allowed to accumulate in the soil, it will limit the growth of the crop. To avoid this, irrigated fields must be periodically flushed with water to remove the salt, and when this 'wash water' is returned to the streams, it increases their salinity
3. *Pollution:* Although irrigation projects add valuable water to fields, there is always the danger of getting "too much of a good thing". Water-logging (excess water) can damage a field as much as drought, so farmers must have a means to remove water from the field when necessary. When this occurs, the water flowing out of the field often carries fertilizers and pesticides that pollute the water into which it flows.

Groundwater

The term "groundwater" often conjures up visions of vast pools or streams of water under the surface. This is rarely the case. Generally groundwater only fills the pore spaces between the grains in underground rock and soil. The ground holds water in much the same way that a sponge does. As surface water soaks into the ground it moves downward through these pose spaces until it reaches an impermeable barrier. While the water is moving downward, it fills some of the available pore spaces, and air fills others. The soil or rock through which the water is passing is wet, but generally it is not saturated with water. As the water begins to accumulate at a barrier, it fills all of the available pore spaces in the surrounding rock or soil. At this point the rock or soil is saturated with water.

The line drawn between the saturated zone and the unsaturated zone, is the water table. Water in the zone of saturation below the water table is known as groundwater. The location of the water table is determined by the amount of recharge (the water put into the ground by precipitation), and the amount of water removed from the ground by natural flow or by human use. The water table is usually not a straight line either. It generally conforms to the topography of the land. The water table may also intersect the ground surface and create a lake, stream, spring, or wetland.

In South Carolina, overall, a good balance is maintained between recharge and removal of groundwater. In most years, the amount of water that is used is about the same as the amount of water that is put back. In other parts of the country, however, the balance is quite different. The large and important Ogallala Aquifer in the central part of the U.S. provides a good example.

Ogallala Aquifer

The Ogallala Aquifer sustains the largest and most productive agricultural region in the US. It is 174,000 square miles large and lies under the states of Nebraska, Kansas, Oklahoma, Colorado, and Texas. It contains as much water as Lake Huron. In 1930 the saturated zone of the aquifer was 60 feet thick, today it has decreased to 9 feet, and in some places under Texas and Nebraska, the drop has been even greater. These decreases are attributed primary to irrigation of farm crops. More than 14 million acres are irrigated by the aquifer, but at the present rate of usage, the supply will only last a few more decades. The loss of this water will pose major problems for the area's agriculture that will either have to change its crops or find other sources of water. In certain areas of the country, however, the depletion of an aquifer can have even more damaging effects. A depleted aquifer can collapse leading to significant subsidence of the land. In coastal areas, the removal of fresh ground water can allow salt water intrusion in its place.

Land Subsidence

Another consequence of excessive use of ground water is ground subsidence. If the rocks or soil that composed the aquifer are no longer saturated with water, they may become compressed by the weight of the overlying rocks. As the rock and soil below the surface collapse, their porosity and permeability are reduced, and the surface itself may subside. This has caused as much as 30 feet of subsidence in the San

Joaquin Valley of California and several feet of subsidence in the Houston and Galveston areas of Texas. Subsidence causes damage to structures in inland areas, but in coastal areas it can cause flooding. Venice, Italy is an example of a coastal area that is drowning due to a combination of causes including surface subsidence due to extensive ground water use.

Salt Water Intrusion

Excessive use of ground water in or near coastal areas can cause another problem beside subsidence. If the water table drops below sea level, saltwater will intrude into the fresh water causing it to become brackish. If the problem is not corrected quickly, the aquifer may be rendered unusable. In South Carolina this problem has occurred in areas along the coast where groundwater is used extensively to irrigate golf courses. In some areas, new pipe lines are being installed to bring water to the coast from further inland.

Water Supply

It is possible to make a rough estimate of the amount of water the continental U.S. receives from precipitation each year. Obviously, the amount of precipitation varies greatly from one area to the next, but the average precipitation for each area is known. The size of each area is known as well, and so it is possible to calculate the average amount of water the continent collects each year. This figure is approximately 1.4 × 1015 gallons. In this form, the figure is not much more than a very large number, but it can be divided by 365 to find out the average amount of precipitation each day, and this figure can be divided by 260,000,000 (the approximate population of the country) to find out the amount of water available for each person each day. That calculation indicates that this country receives about 15,000 gallons of precipitation per person per day. This is a smaller number, but the number still does not have a great deal of meaning. Does it indicate that this country's supply greatly exceeds its demand? Is the supply about equal to demand? Is the supply significantly less than demand? There is really no way to tell until something is known about the average American's demand for water installed to bring water to the coast from further inland.

Demand

Human water consumption is great. We use some water for 'domestic' purposes such as: drinking, bathing, washing clothes and dishes, flushing toilets, and watering lawns and gardens. Altogether each person in a North American home uses approximately 90 gallons of water each day for such purposes. Only a very small percent of this amount is used for drinking. Most of the rest is used as a means to transport waste materials.

Water is also used for "agricultural" purposes, primarily irrigation. In the U.S. about 670 gallons is used per person per day to grow the crops we use.

Finally, this country uses a great deal of water for 'commercial and industrial' purposes. These include water used for cooling of all sorts, for cleaning, for transporting material, and as a component in manufactured materials. On the average, commerce and industry use 1,200 gallons per day for each person in the country. If the figures for domestic, agricultural, commercial, and industrial use are combined; the total is just under 2,000 gallons per person per day. If this very rough figure representing the demand for water, is compared to the equally rough estimate of this country's supply of water, the result shows that we receive from precipitation over 7 times as much water as we use. If this is the case, why does this country have any problems with water? Consider a couple of possible answers to this question, and then click here to see some answers.

Distribution and Pollution

One of the problems is Distribution. Water is not in the place where it is needed, or it does not arrive at the time when it can be used. For example, one portion of a state might receive a great deal of rain while the other portion receives almost none. This is the case in Washington State which receives abundant rain along the west coast, but experiences near desert conditions in the east. Similarly, an area may receive 50 inches of rain per year (about the amount received annually in central South Carolina), but if all of that precipitation comes in the winter, there may be severe shortages of water in the summer. In both of these cases, better distribution of the water would solve many of the problems. Flooding is also a distribution problem in that some areas receive more water than they can handle.

Another problem is Pollution. There may be an abundance of water, but if it is too polluted to use, the problem is the same as having no water at all. The solution, in this case, is to determine the source(s) of the pollution, and consider ways to prevent it.

Improve Distribution

Humans have devised a great number of ways to deal with the distribution of water. The most common solutions involve canals, dams, and desalination plants; but there are also some less traditional systems such as the use of ice bergs and cloud seeding that should be considered.

Canals: For thousands of years humans have dug canals and built aqueducts to carry water from one place to another. The All American Canal in southern California is a present day example of a simple solution to a distribution problem. Large portions of southern and central California have a warm and dry climate that many farmers and suburbanites find nearly ideal. The only quantity lacking is water. To solve this problem over the past century, California has constructed a large number of canals and aqueducts that take water from river such as the Colorado and the Feather and bring it south and west to the centers of population and agriculture. The projects have been well planned and they successfully supply abundant fresh water where it is needed. However, it is important to remember that such projects have a monetary as well as an environmental cost.

Dams-Like beavers, humans have used dams for centuries to control their supplies of water. Many dams have two or three functions:

1. *Flood Control:* A dam across a river with a sizable lake behind it can be used to control the flow of the river. Water accumulates in the lake during times of heavy precipitation, and water is released from the dam to maintain the flow of the river when the weather is dryer.

2. *Irrigation:* Water that accumulates behind a dam can flow or be pumped into a series of canals that irrigate surrounding farmland. Simultaneously, water from the lake will seep into the ground and raise the local water table. Along with these and other advantages, it is also necessary to consider the environmental disadvantages of a dam. It certainly disrupts the natural flow of the river; the lake it creates floods and alters a large amount of land; and the lake's large surface area looses a lot of water to evaporation in dry climates.

Desalination Plants: There are several ways to remove the salt from sea water and make it drinkable. The water can be made to evaporate and the vapor recondensed into water, the salt water can be forced through a membrane that only allows water molecules to pass, or the water could be frozen (the salt is forced out of the water as it freezes) and remelted to obtain fresh water. However it is done, desalination is much more expensive than simply pumping water out the ground or out of a stream. Key West in Florida is the only American city that runs exclusively on desalinated water, and the cost is about ten times as great as for more traditional sources. However, abundant fresh water can be well worth the cost if there is no other way to get it.

Icebergs: A process that has been proposed many times, but never used to a significant extent, is to obtain fresh water by capturing and melting icebergs. Most icebergs are large pieces of glacial ice that have fallen into the ocean. If these ice giants could be captured and towed back to an enclosure in a warm and dry area, they could be allowed to melt and produce fresh water for local use. This idea was proposed at one time for southern California, and it was suggested more recently for cities in the Middle East, but it has never been successfully put into practice.

Cloud Seeding: In theory, it is possible to produce rain on demand if the conditions are correct. Rain occurs naturally when the air becomes saturated with water and water vapor condenses on minute airborne particles. When the water-covered particles become sufficiently large, they fall as rain. Therefore, it should be possible to produce rain by adding minute particles to moist air thereby providing 'nuclei' on which rain drops could form. Particles of dry ice (frozen carbon dioxide) and crystal of silver iodide have been used, but the process has never been very reliable.

Flooding

The cause of some floods seems to be obvious; a river that overflows it banks, or a rain storm so intense that the water cannot soak into the ground or flow away. The solutions to these problems seem obvious as well; build dikes or levees along rivers and dig canals to drain away rain water. An even simpler solution would be for people not to live in places where floods are apt to occur. However, there are problems with all of these solutions. First of all, many people live in flood-prone areas because these areas have rich soils and abundant moisture deposited by past floods and thus they produce good crops. Second of all, measures such as dikes and drainage canals that force the water to go elsewhere, often aggravate flood problems in other areas. As with many other problems, flooding is best handled by working with nature rather than against it. Rivers overflow their banks when they receive more water than they can handle. In some cases this is unavoidable such as during heavy rain or rapid melting of snow, but other times it is produced by human actions. Levees that do not allow a river to spill its excess onto a flood plain in one area, are going to increase the possibility of flood in another. Canals that drain urban areas rapidly, significantly increase the flow of local rivers. To lessen these problems, it may be best to remove residential areas from flood plains and allow rivers to flow through these areas as they always did during times of high water. Similarly, in cities where buildings and pavement prevent water from soaking into the ground as it used to, it makes sense to build "ponds" or "lagoons" where rainwater can be held until it has soaked into the ground or until it can be released more slowly into local streams.

NATURAL RESOURCES

Fossil Fuel

Today's economy is based on an abundant supply of low-cost energy. Nearly 90% of this energy is derived from the fossil fuels: oil, natural gas and coal. Smaller amounts come from uranium and hydroelectric power. Aware of our dependence on energy, we use geological knowledge to search for new sources of these traditional fuels, and at the same time, we seek to develop new sources of energy. In all of these efforts, we must remain aware of the earth's limitations, and try to maintain a delicate balance between our need for energy and our desire to protect the environment.

Coal

Coal supplies about 20% the energy used in the U.S. The fact that plant fossils are found in seams of coal is a clear indication that coal forms from the preservation and alteration of plant material. Swamps appear to be the ideal location for the origin of coal because they produce luxuriant plant growth and provide a shallow-water environment in which organic material can be preserved. Shallow water that is warm and stagnant contains very little oxygen, so the leaves and stems that fall into it are sealed off from scavengers and thus preserved. These materials are first converted into peat, a low-grade fuel that contains about 50 percent carbon.

If the peat remains buried and is subjected to elevated temperatures and pressures, some of its volatile components are driven off. Since carbon is not lost in this process, its relative percentage is increased and the material becomes more compact. The process proceeds to lignite with about 70 percent carbon, subbituminous and bituminous (soft) coal with about 80 percent carbon, and finally anthracite (hard) coal with over 90 percent carbon.

Since coal represents the accumulation of material on the floor of ancient swamps, the coal tends to form in layers known as seams. Seams can range in thickness from a few inches to many feet. Usually only seams that are several feet thick are mined.

Mining

About 70% of the coal mined in the U.S. comes from Open Pit or Surface mines. If a coal seam is relatively close to the surface, it can be mined by removing the soil or rock above until the coal is exposed. The coal can then be removed using enormous power shovels.

This type of mining is efficient since it can employ large pieces of equipment which are operated by relatively few miners. Open pit mines are also reasonably safe since they are not likely to collapse or explode, and they do not subject the miners to polluted air. On the other hand, mines like this are much more likely to damage the local environment due to erosion and the production of acid mine drainage.

About one third of the coal mined in the U.S. comes from under ground or tunnel mines. When seams of coal are deep below the surface, it is necessary to dig a vertical shaft down to the coal deposit and then mine it by digging horizontal tunnels called adits out away from the shaft. This type of mining is less likely to damage the local environment, but it is more costly. The need to use smaller equipment and a larger staff of miners means that underground mines are more expensive to operate. These mines are also likely to be more dangerous for the miners since they can collapse and they can trap explosive gasses and dangerously dirty air.

Petroleum

Petroleum supplies about 40% of the energy used in the U.S. Petroleum forms from the preserved remains of plants, bacteria, algae, and other microorganisms that lived in the ocean and were buried in ocean-bottom sediments. For preservation to be successful, these organic remains must be isolated from scavengers by rapid burial, or by deposition in water with a low supply of oxygen.

Since marine life is abundant in the shallow waters of continental shelves, and since there is a great deal of sand and clay being deposited in these areas, it is likely that most petroleum formed along continental margins. Once collected, the organic debris is slowly converted to petroleum as temperatures and pressures rise. These conditions convert the organic material into hydrocarbons. When hydrocarbons form, their low density causes them to rise and they move into adjacent permeable rock where they can collect in oil traps. Petroleum is removed from oil traps by drilling.

Hydrocarbons

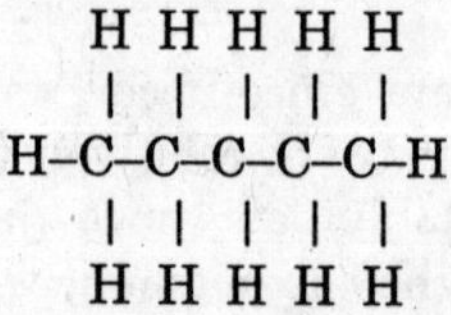

Structural Diagram for Pentane

Hydrocarbons are chains of carbon atoms with hydrogen atoms bonded along the outside and at the ends. The simplest hydrocarbon is CH_4 (Methane). This is the primary component in natural gas. As the chains of carbon atoms get longer, they form liquids such as octane (a chain of eight carbon atoms), and eventually thicker materials, such as tar and asphalt.

Oil Traps

Hydrocarbons collect in reservoirs when their upward migration is halted by an impermeable barrier. Barriers can be faults which cut across the layer which contains petroleum, or places where the layers have been deformed into an arch or dome, which traps petroleum at its highest point.

Drilling

The primary job of the petroleum geologist is to locate potential oil traps and plan a way to drill into the trap and determine its contents. If the trap contains a sufficient quantity of oil, the area is developed into an oil field by drilling several more wells into and around the trap. Some wells allow the petroleum to flow to the surface, others are used to pump water in below the oil and float it to the surface and force it up through the production wells. A properly planned oil field can eventually recover about 50% of the oil in the trap. If the drillers are careful, very little oil is spilled on the ground or allowed to escape into the local water supply.

Natural Gas

Natural gas forms under the same conditions as petroleum, and is the most abundant type of hydrocarbon. The lighter and more fluid gas can move upward easily, until it is caught in a trap or it escapes at the surface. Natural gas is recovered from the ground by drilling.

SOME ALTERNATIVE FOSSIL FUELS

As the conventional sources of petroleum are depleted, one choice is to turn to materials similar to petroleum known as syncrude, or synfuel. Various types of syncrude can be obtained by processing oil shales and tar sand. Oil shales are fine-grained sedimentary rocks containing organic material that can be released as an oil-like substance if the rock is heated. Tar sand, as the name suggests, is sand containing a tarry material which can also release oil when heated. The world's resources of oil shale and tar sand are significantly greater than the reserves of crude oil, and nearly half of these resources are in the United States. Another approach coal gasification, the conversion of coal into a flammable gas. Older processes used coal to produce water gas or blue gas which was composed primarily of hydrogen and carbon monoxide. More modern processes convert coal into methane.

Though promising, all of these fuels present some problems. Currently they are more expensive to produce than oil or coal, and oil shale and tar sand pose a greater threat to the environment since they must be mined and processed with water, and the shale and sand must be disposed of after the oil has been removed.

Relatively Inexpensive: Relative to most other sources of energy, fossil fuels are less expensive.

Fossil Fuel Reserves

Reserves are deposits of a resource which have been located and are know to be minable. Together the United Sates, China and the former Soviet Union hold about 85 percent of the world's coal reserves. The United States, with about 15 percent of the world's coal, could supply its own needs for nearly 400 years at the current rate of use. If it becomes necessary, coal can be made into a liquid form known as syncrude, which could extend our supplies of petroleum. The United States produces about 11 % of the world's oil, but uses about 26%. If we were to depend solely on our own oil resources, our supply would be depleted in about 20 years. Worldwide, reserves of petroleum are probably sufficient to meet the current demand for about 100 years. Technology may provide a partial solution to the problem. Current methods of oil production leave about half of the oil in the ground. Procedures designed to enhance the recovery of oil use steam, hot water, or carbon dioxide to mobilize deposits that were previously inaccessible. However large the world's reserves might be, they represent a set amount which will eventually become depleted if use continues. Eventually it will be necessary to find alternatives to fossil fuels.

Alternatives to Fossil Fuels

As mentioned earlier, the world's current use of crude oil and natural gas is expected to deplete the supplies within the next 100 years. Extensive development of oil shale and tar sand will extend our supply of petroleum, but it will not solve the problem. Increasing our dependence on coal might add another century to our supply of fossil fuel, but the carbon dioxide produced by burning any sort of fuel may turn out to be a more significant limit. Many people believe that the only solution lies in conservation of fossil fuels and the development of alternative sources of energy.

ALTERNATIVE SOURCES OF ENERGY

Calculations indicate that the world's current use of crude oil and natural gas is expected to deplete the supplies within the next 100 years. Extensive development of oil shale and tar sand will extend our supply of petroleum, but it will not solve the problem. Increasing our dependence on coal might add another century to our supply of fossil fuel, but the carbon dioxide produced by burning any sort of fuel may turn out to be a more significant limit. Many people believe that the only solution lies in conservation of fossil fuels and the development of alternative sources of energy.

Nuclear Energy

When most people mention Nuclear Power, they are referring to power from Nuclear Fission. This is the type of nuclear power used to generate electric power and operate navy ships. This is the same process which fueled the atomic bombs used during World War II. This type of energy is obtained from the break-down, or Fission, of atomic nuclei. Another way to obtain energy from atoms is Nuclear Fusion. This process is based on the build-up, or Fusion, of atomic nuclei and is the system which creates energy in the sun, and which fuels the hydrogen bomb.

Nuclear Fission

The nuclear energy first released by the atomic bomb, and later applied to more peaceful purposes, is based on nuclear fission, or the splitting of atomic nuclei. Uranium 235 is one of several fissionable elements--elements whose nuclei can be induced to split, forming two new elements. As the atoms split, a great deal of energy is released. This energy can be used to create steam for electric generating stations or to power naval vessels. The fission of one gram of uranium 235 liberates as much energy as 3 million grams of coal. In the United States 113 nuclear reactors provide about 20 percent of our electric power. Worldwide, more than 400 nuclear reactors produce electric power in 25 countries.

Nuclear Fuel

Uranium is a rare element (only .00016 percent of the earth's crust), and the problem is compounded by the fact that less than 1 percent of all uranium is fissionable. Uranium is extracted from uraninite (called pitchblende), which is a uranium oxide deposited in igneous and sedimentary rocks. Some of the richest deposits are found on the Colorado Plateau, and in adjacent parts of Utah, Arizona and New Mexico. Canada, Australia, South Africa and Brazil also have significant deposits.

It is possible that the uranium from demilitarized weapons will also become an important commercial source. At one time nuclear power was considered a cheap and clean alternative to fossil fuels, but the cost of building and operating safe reactors, and the difficulty of finding sufficient fuel have become significant problems. An additional problem is the fact that we have not yet found a safe and acceptable way to dispose of nuclear waste.

Nuclear Waste

The process of nuclear fission creates a number of highly radioactive elements which are not themselves useful for additional fission, but which obstruct the fission of uranium. As a result, nuclear fuel eventually becomes "polluted" with these wastes and must be replaced. When nuclear power was first envisioned, planners expected to remove the polluted fuel, clean out the contaminants, replenish the used fuel, and replace the fuel containers in reactors.

This operation is known as "Reprocessing". Reprocessing is used in several countries outside of the U.S., but it is considered too risky to be employed here. As a result, used nuclear fuel containers must simply be stored. The highly radioactive "wastes" they contain cannot be neutralized and therefore they must be permanently sealed off from the environment. At present, most such containers are stored in "Spent Fuel Pools" (large shielded tanks of water) at the sites where the fuel was used. Eventually these wastes will need to be reprocessed in some way, or sealed away for thousands of years until they decay to safe levels. The U.S. government has not yet decided how to handle this waste. A likely possibility is that it will be mixed with molten glass, placed in durable containers, and eventually buried in mines. One such mine is being constructed and tested at Yucca Mountain in Nevada.

Nuclear Fusion

Nuclear fusion is the opposite of fission. In this process, the nuclei of smaller atoms are forced together to form larger atoms. The simplest way to look at this is the joining together of four hydrogen atoms, each of which has one proton in its nucleus, to form a helium atom which has a nucleus with two protons and two neutrons. When these atoms are joined, the product weighs slightly less than the components. The small amount of mass which is lost, is converted into a vast amount of energy. In a bomb, this energy is the weapon's "blast". The problem is to harness this energy so that it can be released in a slower and more controllable manner. As yet, this problem has not been solved. An additional problem is the fact that a great amount of heat and pressure is required to initiate a fusion reaction. This is why the now discredited idea of "Cold Fusion" (fusion at low temperatures and pressures) appeared to be so attractive several years ago.

Geothermal Energy

We know that the earth is hotter on the inside than on the outside. Most of this heat is produced by the decay of radioactive materials, and it provides the energy to build mountains, create volcanoes and produce earthquakes. We can gain access to this heat when water passes though heated rock and rises toward the surface. The most abundant geothermal energy is supplied by "low-temperature" water between 80° C and 180° C. This water can be used to heat homes and industrial facilities. Geothermal sources that can be used to generate electricity must be hotter than this, and they generally occur as natural sources of steam in recent volcanic areas.

The world's largest natural supply of steam is The Geysers, north of San Francisco, where steam is harnessed to supply about half of the city's electric power. Ironically, a source of power that appears to be clean and safe can pose an environmental threat. The ascending water or steam can pick up a heavy load of salts and toxic materials from the surrounding rocks, and thus create a distinct disposal problem. In addition, the local terrain may begin to subside if a great deal of water or steam is extracted and not replaced. Technically, the earth's supply of heat is inexhaustible. In reality, the heat that is easily accessible is in hot bodies of rock close to the surface. When these have been cooled, it will be necessary to drill deeper and use more expensive equipment to find new sources.

Solar Energy

The sun can supply all of the energy we need. Solar energy is clean and nondepletable, but it will require some new technology before we can use solar energy efficiently and inexpensively on a large scale. Presently our use of solar energy can be divided into two primary categories. One category includes Passive Methods, such as the design of buildings, and the other category is composed of Active Methods, such as solar collectors and batteries.

Passive Methods

Passive use of solar energy involves designing buildings so that they collect and store heat when it is needed, and avoid heat when that is appropriate. This can be achieved through the placement of windows and the selection of building materials. For example, windows might be arranged such that winter sunlight enters and warms walls and floors made of concrete and brick. The heat stored in the masonry material is then radiated to keep the room warm after sunset. The same windows might be protected by overhangs which block sunlight during the summer when the sun is higher in the sky. A well designed passive system can supply nearly all of the energy needed to heat a house. A solar heated house in Canada built by the University of Saskatchewan has an annual fuel bill of $40 which can be compared to $1400 for the average home in that area.

Active Methods

Humans have designed a great variety of mechanisms to harness solar energy, which range from ancient windmills to sophisticated solar batteries. These are vastly different ways to collect and use solar energy. The battery is a mechanism which uses solar energy directly. The energy from the sun is used as it arrives. The windmill also uses solar energy, but it does so indirectly. The sun's energy is stored or transformed in some way before it is used.

Direct Sources

Most solar collectors are insulated boxes mounted on the roofs of buildings. Sun light enters the boxes through layers of glass and heats the interior. The heat is then carried out of the collector by moving air through the box or pumping water through a series of pipes inside it. The heat is usually stored in the building below as hot water until it is needed as domestic hot water, or as a source of heat for the building. In Israel and some surrounding countries, nearly half of the domestic hot water is produced by such systems.

Solar batteries, or photovoltaic cells, contain silicon or other materials which emit electrons when struck by light. A flow of these electrons becomes an electric current. Presently these cells are expensive and not very efficient, so electricity produced in this way is three or four times more expensive than electricity from a coal-fired generating plant. Even so, photovoltaic cells are an important source of electric power in remote areas and for navigation buoys and traffic signs, which cannot be easily connected to an existing source of electric power. As research improves efficiency and increased production brings down the cost, photovoltaic cells will become more acceptable as clean and dependable sources of electric power.

Indirect Sources

Solar energy can be harnessed indirectly in several ways. One way is to use biomass, material of biologic origin, such as wood or agricultural waste, that can be burned or converted to a liquid or gaseous fuel. When compared to coal or natural gas, biomass might be considered "unfossilized fuel". In either event, it is questionable if crops of any sort should be grown as a supply of energy. One problem is the environmental threat from the use of water, fertilizer, and pesticides. Another dilemma is whether it is acceptable to use arable land to supply fuel when the world is having trouble supplying enough food for its growing population.

The production of hydro-electric energy is another indirect application of solar energy. The sun powers evaporation and drives the circulation of the atmosphere, the resulting rain is collected in lakes and runs through turbines to generate electric power. This is clearly a clean, safe and cheap form of energy, but its expansion is limited by the fact that considerable land would be lost to the lakes created by dams, and many scenic areas could be spoiled. A final possibility is wind power, which is also derived from the sun, and is clean and safe. Overall the prospects for solar energy look good in the future. It has the potential to be a safe, clean and economical way to meet the world's increasing demand for energy.

Conservation

Using energy more efficiently is equivalent to finding new sources. The rise in the price of oil in 1973 initiated an effort to conserve that has saved four times more energy than has been developed from all new sources since that time. There is no reason why these conservation efforts cannot continue. The necessary technologies are generally simple, requiring such things as a change from incandescent to fluorescent lighting; the installation of better home insulation; the construction of more efficient appliances; and the design of better automobile engines. In almost all cases the cost of conservation is less than the cost of the energy it will save. For example, the cost of modifying appliances to be more efficient can cost 1 or 2 cents per kilowatt-hour of energy saved, while generation of electric power from coal-fired plants cost about 6 or 7 cents per kilowatt-hour.

MINERALS

Mineral resources are the materials derived from the Earth's surface which can either be single mineral products or rock products. Minerals, by definition, are naturally occurring, inorganic crystalline solids with a set chemical composition. Examples of economically desirable minerals would be gold or silver, bauxite (aluminum source), feldspar, quartz, and diamond to name just a few. Rock products are also considered in this section. Rocks are mixtures of different minerals. Economically important examples of these include sands and gravels, crushed stone, dimension and facing stone. Fossil fuels are not included here because of their organic origin, and because our intense dependence on them means they deserve special consideration on their own. Mineral resources form the foundation of our industrial society and an understanding of their origin, uses, mining processes and availability is essential to making informed decisions about them.

ORES

An ore is a body of rock containing a valuable mineral which is sufficiently concentrated to be mined at a profit. The higher the concentration of a mineral in the ore, the less expensive it is to remove. If the market value of a mineral is high, companies can afford to go to greater trouble to extract the material, as it is then worth the labor and cost to mine rock with a lower concentration of the mineral. If prices drop, they cannot afford the extra cost involved in extracting the material. Another factor that affects whether or not the mineral will be mined is the technology available to extract it. As mining technology advances, some previously unavailable ores become minable. Gold, for example, can now be chemically leached out of rocks that would not have been considered ores in the past.

Uses

These mineral and rock products are used for a huge variety of purposes. If the things we use are not grown as a crop, they must be mined in some fashion from the Earth's surface. Iron, aluminum and copper are used in large quantities for packaging, transportation, construction, and metal fabrication. Nonmetal resources include salt, clays used in paint, rubber and paper coating, quartz used in fiberglass and optical glass, and phosphate used in fertilizers. Sand, gravel, and quarried stones are used in road and building construction. The tables in the nonrenewable section show the quantities of these materials that we use.

Teaching Tip: Ask the students to keep track of the materials they use in a day, and make a list of the rocks or minerals that were used to produce them.

Formation in Rocks

Mineral deposits occur in a variety of geologic settings. Many form in association with the igneous rocks which arise from cooling magma. As magma begins to cool in a magma chamber (the space below the Earth's surface filled with magma), some crystals solidify before others. If the crystals have a high density (like metals), they will settle to the bottom of the chamber. Over time this can produce a great accumulation of material. Chromite and magnetite often occurs this way. Magmas with different compositions, produce different minerals when they cool. Feldspars, used in ceramics, form this way as do tourmaline and beryl. Metamorphic processes (which involve applying heat and pressure to a rock without melting it) produce useful minerals such as graphite and asbestos. Both igneous and metamorphic processes are involved in the formation of important hydrothermal ore deposits. Hot fluids can dissolve, transport, and redeposit metals. The Mid-Ocean Ridges are a location where this process can be seen occurring today. At these spreading centers, hot rocks are moving upward toward the Earth's surface. Ocean water circulates through these hot rocks, dissolving minerals as it is heated, and redepositing them as it cools. Many metal deposits are formed this way. The importance of heat in the formation of these deposits helps geologists narrow down the areas in which to search for them. Heat is generated at plate boundaries, and using knowledge about the current and ancient location of plate boundaries, they can concentrate their efforts on areas with the highest probability of containing the desired mineral resources.

Some deposits are formed, or concentrated, by sedimentary processes which occur at much lower temperatures than the igneous or metamorphic processes. Salt (known by the mineral name halite) and gypsum form from the evaporation of briny water. Major iron ore deposits, known as banded iron formations, are believed to have formed from precipitation on the ocean floor. Earth's primitive atmosphere lacked oxygen. As oceanic organisms began producing oxygen, it caused the precipitation of iron, which had been dissolved in the sea water. While many mineral formation processes occur today (albeit at slow rates), this is one type which cannot occur today at any rate. Placer deposits, usually of gold, diamonds and tin oxides, are formed by water, which sorts materials according to their density. Streams flow at different rates in different places. When the stream's rate of flow decreases, it may deposit dense minerals which it no longer has the energy to transport.

MINING

From time immemorial, humans have derived resources from the Earth's surface. It has always been a difficult process. Decisions about what to mine, when to mine and how to mine are related to several factors: the demand for the materials, the ways in which wastes can be handled, and information about which materials can be recycled. People who make these decisions must become familiar with these and many other factors in order to make wise decisions.

2

Resource Requirement

Taking a survey of the materials you use during the day can reveal just how dependent we are on the minerals and rocks mined from the Earth.

We have already discussed how much of these resources we use in the section on Non-Renewable Resources. All the metals we use in plumbing, automobiles, ornaments, photographic film and construction are derived from the Earth. All the rock products used in the construction of roads and buildings are mined. Today, plastics substitute for many objects that were formerly made from metal, and although plastics are not usually included as mined substances, their raw material is petroleum, which is derived from the Earth.

Environmental Consequences

Although the materials mined from the Earth are useful to us we must recognize the environmental consequences of their removal. The different mining processes produce their own individual problems. These include land collapse, sediment pollution of water bodies, and the leaching of pollutants from tailings. Most mines also produce some degree of dust and noise pollution. Open-pit and strip-mines are also responsible for habitat destruction.

Underground Mines

The type of mining which is employed depends on the shape and orientation of the rock counting the ore. Underground mines are used to reach bodies of ore which are a long way from the surface. These mines are less obtrusive than open-pit or strip mines. The process occurs beneath the surface, and other than shaft outlets, truck traffic and occasional tailing piles, it goes on barely detected.

However, one great risk of underground mines is the collapse of mines when erosion from ground water enlarges the shafts or weakens the materials which support the mine's roof. Not only is this a risk to miners, but it can also cause substantial landslides at the surface.

Open-Pit Mines

Open-pit mines are used for massive ore bodies, which lie close to the surface. These mines change the land surface extensively, generally permanently, as nearly all of the material removed from the pit is valuable. One possible use for the land after mining is as a landfill site. One of the most significant environmental problems with these mines is the weathering of the exposed rock, which can produce contaminants in surface and ground water.

Strip-Mines

Strip mines are used to gain access to ores which are near the surface and oriented horizontally, such as beds of coal. All surface vegetation, soil and rock are striped away and piled nearby in spoil banks. Though reasonably safe and inexpensive to operate, the problems associated with these mines are numerous. They seriously disrupt large areas of land, and their associated spoil banks erode easily and can pollute local bodies of water with sediment and other contaminants. The mines can also change topography significantly and disrupt local drainage patterns. Today, strip mines are under strict regulation, and the companies which operate them must reclaim the affected areas by grading banks and replacing vegetation.

NONRENEWABLE RESOURCES

Mineral resources are considered nonrenewable resources. Although some of these minerals are being formed today, they are being consumed at such a fast rate that new formation could not possibly keep pace with the rate of extraction. The huge quantities of these materials which are used.

A proven reserve is a mineral deposit which can be profitably mined with current technology. The deposit has been studied and quantified as to its location, size and quantity. Sub-economic resources have been discovered, but would not be profitable at current prices, or could not be extracted with existing technology. This distinction is important when the question of "how long will a mineral resource last?" is considered. Another factor which must be considered in determining the life span of a resource is the expected demand.

The expected demand is a function of the population and the per capita consumption of the individuals. In most industrialized countries, the population grows relatively slowly, so people believe population growth is not a problem. However, the quantities of resources they use is disproportionately high and continues to rise, so that each individual takes a great toll on the environment.

The graph here demonstrates this escalating per capita use of resources. World population is increasing, and as standards of living continue to rise, per capita use of resources will increase as well. Since the exact rate of increase is impossible to predict, many assumptions are made when predictions about the life span of a resource is made. Notice the assumptions made on the graph here of expected life spans of mineral resources.

Some people believe that limited mineral resources do not pose a real problem, because as quantities dwindle, the prices will increase, making previously unminable resources profitable. Substitute materials will also be developed to replace increasingly expensive ones. However, there are environmental costs associated with any increase in mining, and this would be particularly true when extracting minerals from low-grade ores, because more land would be tampered with, and a higher proportion of waste rock would be produced. It is much more logical to consider reductions in consumption and increases in recycling as the primary methods to deal with the world's increasing demand for mineral resources.

Global considerations aside, it is important to consider this country's high dependence on imports. The United States' civilian and defense industries rely heavily on imported minerals. When shortages of these minerals occurred during World War I, the government initiated a program of stockpiling strategic minerals. This program continues today, and the United States has a three-year supply of many critical minerals in the event that their flow is interrupted due to political or economic causes.

LAND USE-FEDERAL PUBLIC LANDS

Approximately 42% of this nation's land has been set aside for public use and enjoyment, and for wildlife--more than any other country in the world. Approximately one-third of the nation's land is managed by the federal government. While federal lands technically belong to all people, many are located in remote areas and are not easily accessible. Approximately 73% of the federal public lands are located in Alaska, and another 22% are located in the western states of the "lower 48".

Components of federal public lands are managed and administered by different federal agencies. The uses to which these lands can be put varies widely with the categorization of the land and with the agency responsible for its management and administration. As a result, each category of public land is susceptible to unique human impacts.

In this lesson, we will examine four categories of public lands, including reasons why the category was established and how each is administered and managed, the allowed and disallowed uses to which each can be put, and any problems which might be besetting them. The four categories of public lands to be examined include:

SYSTEM OF THE NATIONAL PARK

Even though prior to 1870 the now famous Yellowstone area of Wyoming, Idaho and Montana had not been really explored, rumors of an area containing black glass cliffs, hot springs from which spewed boiling water and smoking hills had been circulating for years. Most of these stories were met with skepticism at best, ranging to non-belief at worst. However, in 1870, after years of rumors, a party of men from Montana decided to explore the land to determine once and for all if the stories that had been circulating for all those years contained even an element of truth. What they found, after exploring the region for about two weeks, was one of the most spectacular geologic phenomena ever seen, anywhere. A series of further explorations yielded not just written descriptions of the area, but also paintings and photographs. Interest ran so high in the establishment of a "natural park" for all people to enjoy and visit, that when a bill was introduced to both houses of Congress in late 1871, it took less than three months before being signed by President Grant. By 1890, three more parks were added to the system, and by 1899, a fifth park had been set aside, all for their scientific and historic significance.

It was not until 1916 that the National Park Service was established as a bureau within the federal Department of Interior. Congress gave the new agency the following mandates:

- The national parks are to be maintained in absolutely unimpaired form for the use of future generations.
- The national parks are to be set apart for the use, observation, health and pleasure of the people.
- The national interest must dictate all decisions affecting public or private enterprise within the parks.

Currently, there are more than 370 areas, comprising more than 80 million acres (approximately 5% of the total land area), that are administered by the National Park Service. In addition to true National Parks, there are several other categories of land areas administered by the National Park Service. These other areas include national monuments, national historic parks, national seashores and national recreation areas. However, most are located in the western United States, distant from urban areas. As a result, many people in urban areas, and many low-or limited-income people from all across the United States have never been to a National Park. As mentioned in the introduction to this lesson, the uses to which national parks can be put are dictated by the Department of Interior and the National Park Service. These include:

- Camping
- Hiking
- Sport fishing

In addition, on-road vehicles and commercial vendors are allowed in the national parks.

Among the uses to which the national parks cannot be put are:

- Hunting
- Timbering
- Mining
- Oil and gas drilling

Furthermore, off-road vehicles are not allowed.

It should be noted, however, that national recreation areas allow some sport hunting, some mining and some oil and gas drilling.

While the national parks and recreation areas are by any measure a huge success and something about which Americans can be justly proud, in recent years, the national parks and recreation areas have been subjected to increasing stress. Much of that stress comes from an ever-increasing pressure by private interests to engage in activities which can turn them a handsome profit--logging, mining, oil and gas drilling, for example. There have even been attempts to completely privatize the national parks, thereby eliminating any "regulation" as to their use.

However, the biggest stress being experienced by the national parks, is in fact a result of their ability to attract people. Today, more than 350 million people annually visit the national parks, and while the number of visitors to the national parks is continuing to increase, the number of park rangers has remained essentially the same and the budgets, in terms of inflation, have actually decreased. These numbers, however, do not tell the whole story, because the number of visitors to specific national parks is nowhere equivalent. The most popular of the national parks, including, but not limited to, the Great Smoky Mountains, Yosemite, Yellowstone and Grand Canyon, attract approximately 80% of the national park visitors. The effect of all these visitors it that we are "loving our parks to death". With an ever-increasing number of visitors comes demands for more accommodations, more roads, and more infrastructure services (water, sewage, garbage removal, etc.).

There is more litter and more crime and there are more cars and therefore more air pollution. Simultaneously, we are seeing a decrease in much of the wildlife population and some of the endangered plant and animal species which inhabit the parks. In the peak season at Yosemite, you are not allowed to take your car unencumbered through the park-you must take a tourist bus. In the peak season at the Grand Canyon, so many planes and helicopters fly over the canyon, that noise pollution is becoming a serious problem. In the Great Smoky Mountains, you may see a black bear on the highway (which is becoming very rare), and traffic tie-ups cause cars to line up a dead stop for miles in both directions.

What can be done to protect these "crown jewels" of America? Several proposals have been put forth, mostly by conservation and environmental organizations. While these vary depending of who draws up the proposals, they generally include:

- Restricting commercial activity-including eliminating or dramatically scaling back commercial ventures within the parks, including logging, mining, drilling, but also including restaurants, lodges and stores
- Reducing crowdedness-some suggestions call for limiting the number of visitor permits during peak season
- Controlling traffic congestion-barring or dramatically curtailing the use of private vehicles, at least during peak season
- Increasing the number of parks and/or increasing park budgets

As you might expect, most of these recommendations are met with strong objections by various private interests.

NATIONAL FOREST SYSTEM

The Forest Service is a bureau within the Department of Agriculture. It was established in 1905, to provide quality water and timber for the Nation's benefit. Over the years, the role of the Forest Service has expanded, in response to public needs, to include management for multiple uses and benefits and for the sustained yield of renewable resources, such as water, forage, wildlife, wood and recreation. Multiple use means managing resources under the best combination of uses to benefit the American people while ensuring the productivity of the land and protecting the quality of the environment.

The Forest Service manages public lands, collectively known as the National Forest System. Currently, there are 155 national forests and 20 grasslands in 44 states, Puerto Rico and Virgin Islands, which encompass 191 million acres, approximately 8.5% of the total land area of the United States.

The main activities of the Forest Service include:

- Protection and management of natural resources-on National Forest Service Lands
- Research-on all aspects of forestry, range land management and forest resource utilization
- Community assistance and cooperation-with State and local governments, forest industries, and private landowners to help protect and manage non-Federal forest and associated range and watershed lands to improve conditions in rural areas
- International assistance-in formulating policy and coordinating U.S. support for the protection and sound management of the world's forest resources.

The phrase "multiple use" is very appropriate to the mission and role of Forest Service. People are allowed a wide variety of recreational activities on national forest lands, including:

- Camping
- Hiking
- Fishing
- Hunting
- Using off-road vehicles-in designated areas

Additionally, other activities of a more commercial nature are permitted, including:

- Timbering-by selective cutting and clear cutting
- Mining
- Grazing

Timbering

While there are several methods used for harvesting trees, the two most widely used are:

- Clear-cutting, and
- Selective cutting

Clear-cutting involves the removal of all trees from a single area in a single cutting. After clear-cutting, the area is generally reseeded, either with a single type of tree or with a combination of trees. Clear-cutting increases the timber (and therefore monetary) yield in a given area, permits reforestation with specific types of trees, and reduces the time required for establishing a new stand of trees. Clear-cutting, however, produces large scarred areas, and in so doing, reduces the use of the area for recreation and fragments wildlife habitats. If clear-cutting reduces biodiversity, it leaves the area susceptible to invasion of destructive organisms. Finally, clear-cutting leads to soil erosion, sediment pollution of surface waters and flooding.

Selective cutting involves the removal of intermediate-aged or mature trees, cut singly or in small groups, creating only small gaps in the forest. Selective cutting is used to reduce crowding and to encourage the growth of smaller trees. It maintains a diverse stand of trees, and does not severely exacerbate soil erosion. Many industry and government foresters maintain that selective cutting is not profitable, and does not promote the growth of the most sought-after trees. However, other private and government foresters have demonstrated that satisfactory profits can be attained through selective cutting.

The National Wilderness Preservation System

Wilderness areas are those where humans are only visitors and are not allowed to remain. These are areas, each generally in excess of 5000 acres, which have retained their pristine character. In 1964, Congress passed the Wilderness Act, "...in order to assure that an increasing population, accompanied by an expanding settlement and growing mechanization, does not occupy and modify all areas within the United States and its possessions...". Additionally, it was recognized that there was a need for ecosystem studies, and a need for the protection of threatened and endangered species which called for preserving wilderness areas.

Unlike the National Park and National Forest Systems, the National Wilderness Preservation System is administered by a single cabinet-level department. Rather, wilderness lands may be managed by one or any combination of the following bodies:

- Forest Service
- National Parks Service
- Fish and Wildlife Service
- Bureau of Land Management

In 1997, there were 657 designated wilderness areas, which encompass more than 104 million acres, approximately 6% of the total land area of the United States. The largest single manager of wilderness areas is the National Parks Service (41%), followed by the Forest Service (34%) and Fish and Wildlife Service (20%). More than half of the wilderness acreage is located in Alaska.

Designation as a wilderness area allows for only strictly controlled recreational uses. Among the uses to which a wilderness area can be put, by humans, are:

- Hiking
- Camping

Among the many uses to which wilderness areas cannot be put include:

- Vehicular activities
- Timbering
- Mining
- Commercial activities

Further, there can be no permanent or even temporary roads, no motorized equipment, no form of mechanical transport.

The designation of wilderness areas is often a highly controversial act. Wilderness designation takes large amounts of land out of use. For that reason, timbering, mining, ranching, energy and development interests are often strongly opposed to expansion of the wilderness system, wanting to be able to put these areas to more profitable use. Furthermore, many people choose not to visit wilderness areas, because of the restrictions on what they would consider daily necessities (automobiles, electricity, piped in running water, rest rooms, etc.). For that reason, many of those people do not understand the necessity of preserving wild areas and do not support the expansion of the concept.

However, many other people who may not visit wilderness areas still support the need for wilderness areas, feeling that they want to know that these areas exist, even if they, themselves, do not use them. Wilderness areas provide undisturbed habitats for wild plants and animals, promoting biodiversity.

THE NATIONAL WILD AND SCENIC RIVERS SYSTEM

In order to protect rivers and river segments with scenic, recreational, scientific, historical or cultural significance, Congress, in 1968, passed the National Wild and Scenic Rivers Act. These river segments are to be kept free of development, straightening and damming. According to the National Wild and Scenic Rivers Act, these river segments are to be permanently administered by an approved agency or political subdivision of the state or states concerned.

The National Wild and Scenic Rivers Systems actually encompasses three types of protected river segments:

- Wild rivers-those rivers or sections of rivers that are free of impoundments and generally inaccessible except by trail, with watersheds or shorelines essentially primitive and waters unpolluted.
- Scenic rivers-those rivers or sections of rivers that are free of impoundments, with shorelines or watersheds still largely primitive and shorelines largely undeveloped, but accessible in some places by roads or railroads.
- Recreational rivers-those rivers or sections of rivers that are readily accessible by road or railroad, that may

have some development along their shorelines, and that may have undergone some impoundment or even diversion in their past.

The only activities to which these river segments (and their shorelines) can be put include:

- Camping
- Swimming
- Non-motorized boating
- Sport fishing
- Sport hunting

Some new mining claims are also permitted in some areas.

In 1997, there were 150 sections of rivers protected as part of this system, corresponding to approximately 3.5 million miles or 0.3% of the total length of the country's waterways. Many environmental groups are pushing to add as many as 1500 new segments of rivers to the system. This ambitious goal, however, is being strongly opposed by many private interests and some local communities. There is also a movement among environmentalists to encourage the Congress to establish a permanent federal administrative unit to manage the system.

Urban Land Use

Urban areas are here defined as cities or towns with populations of 2500 people or more. The term "degree of urbanization" is used to define the proportion of a nation's population living in urban areas. Worldwide, the number of people living in urban areas has tripled since 1950. Because there is such a dense concentration of humans in these urban areas, there are very significant stresses both on the people living there and on the physical environment. Some of the important topics associated with proportionally high degrees of urbanization are:

Immigration to Urban Areas-opportunities that are driving people into urban areas in most parts of the world;

Urban Stresses-both social stresses and stresses on the environment; and,

Emigration from Urban Areas-the stresses that cause people to leave, as well as the opportunities that draw people away from, urban areas.

HISTORICAL GROWTH OF URBAN AREAS

Many of us have heard about "the threat of urban sprawl"-the thought that people are flocking to urban areas in the U.S. and worldwide at an ever-increasing rate, and that consequently these growing population centers are rapidly consuming agricultural land, parks and forests, and grasslands. To understand the reality of this issue, both now and in the near future, two points need to be considered:

Population Growth in Urban Areas

The population of urban areas is increasing very rapidly. In 1900, only 14% of the world's population lived in urban areas. In 1995 that percentage had grown to 42%, and by the year 2025 experts estimate that 60% of the population of the world will be concentrated in urban areas. The urban population growth rate in developed and in less developed countries.

The growth rate of urban population in less developed countries (LCDs) is exponential. In 1950, only about 19% of the people in less developed countries lived in urban areas. In 1995 the percentage was 34%, and the percentage is projected to be at least 57% by 2025.

The urban population growth rate in the more developed countries (MDCs), including the U.S., is similar to that of LCDs only in that the percentages are still increasing. The growth rate in urban areas in the developed countries is significantly slower, however, than that in the less developed countries. In 1950, the percentage of people living in urban areas in MDCs was approximately 56%. In 1995, the percentage was approximately 73%, and in 2025 the percentage is projected to be 84%. In the U.S. specifically, the percentages of urban inhabitants has grown from 63 % in 1950, to 75 % in 1995 and is expected to be approximately 90% in 2025.

Worldwide growth of urban areas has led to the development of mega-cities, urban areas with 10 million or more people. In

1995, there were 15 such localities, led by Tokyo with over 26 million people. Most mega-cities are located in less developed countries. Only two, New York City and Los Angeles, are found in the United States. By the year 2000, the number of mega-cities is projected to grow to 26, with more than two-thirds located in less developed countries.

Areal Growth of Urban Areas

Much has been written and said about the areal growth of urban areas, often termed "urban sprawl". However, despite near-explosive growth of urban populations throughout the world, the total amount of land occupied by urban areas is still relatively small.

Even in the U.S., the world's most urbanized nation, only 3 % of the total land area is so occupied. So, what is the problem?

The problem resides more than just in the amount of land occupied by urban areas. The bigger problem is the quality of the land that tends to be occupied by urban areas. Many urban areas develop in the most agriculturally valuable land, because it is important to have a central area for farmers to tend to their agricultural business. However, as urban population increases, the area occupied by the people must also increase. Therefore, the area occupied by urban settlements grows at the expense of the farmland that many were founded to serve. This problem is just as important to the developed nations as it is to the developing nations.

In 1994, the government of Taiwan decided to convert 18% of that country's farmland to factories and housing over a period of 10 years. Not only did this action break an ancient social custom; it placed Taiwan in the lead in making the shrinking of agricultural land a matter of official policy. This policy might have made some sense if there was a surplus of Taiwan agricultural products. However, Taiwan was having to import more than 80% of its grain.

Some places are trying to halt or at least slow the spread of human settlements onto agricultural and forested lands, and wildlife habitats. The state of Oregon, long a leader in managing growth, passed legislation in the 1970s to require urban areas to establish urban growth boundaries. This legislation was put into place to protect open space, agricultural land, forests and wildlife habitats. As a result, not only was outward expansion slowed considerably, but also many urban areas became more efficient in terms of mass transit, air quality and waste management.

Immigration to Urban Areas

Some people are pulled to urban areas in search of what they perceive to be a better life, and other are pushed to urban areas by the growth of modern, mechanized agricultural practices that favor large farm complexes over small, more labor-intensive family farms. Overall, there are three main factors that bring about immigration to urban areas:

Perceived Employment Opportunities

Much of the migration to urban areas, both in developed countries and less developed countries, is due to the increasing difficulty of earning a decent living in rural areas. The perception is that employment opportunities are greater in the cities. Most of the rural poor are farmers whose small farms or obsolete farming practices do not allow them to compete with larger, more modern, highly mechanized farming corporations. As these rural poor lose their land, they have little alternative but to move to the cities for employment. Unfortunately, many lack the education or specialized technological training necessary to find good employment. Usually they have no choice but to take low level jobs for very low wages and few, if any, benefits; they are often forced to work very long hours, and too often the work is in unhealthy, unsanitary and/or actually dangerous conditions. They are exposed to increasing levels of urban stress, and find themselves transformed from rural poor to urban poor. As a result, poverty is the largest single growth aspect of cities, especially those in less developed countries.

Increased Social and Cultural Opportunities

One important factor drawing people to urban areas, especially in developed countries, is the enhanced social and cultural opportunities afforded them in cities. Cultural opportunities such as libraries, museums, orchestras, live theater and universities are located almost exclusively in urban areas. Access to these features is no doubt a·factor in the movement of some people to cities. In less developed countries, and to some extent in the developed countries, educational opportunities also pull many people into urban areas.

Enhanced social opportunities exist in urban areas for some people. These include more and varied opportunities to interact with others of similar age and interest-senior citizens can interact with people their own age, young adults can meet and develop networks of friends, and younger citizens can participate in group activities such as sports and social events. In many cases, neighborhoods in inner cities actually function as small towns, and provide their inhabitants with an extended support group.

Availability of Goods and Services

There can be little argument regarding the enhanced availability of goods and services to urban dwellers as opposed to people who live in rural areas. This advantage for people who live in urban areas holds true in both developed and less developed countries, but the gap between urban and rural is greatest in developed countries. Food and water are centralized and more easily available; access to clothing is easier, as is gasoline, furniture, entertainment and bicycle parts.

Similarly, necessary services are also more readily available in urban areas. Access to health and medical care is much easier. Waste disposal is often more centralized and less of a worry to individuals; no need for a weekly trip to the local dump or landfill. In many urban areas, even those with no mass transit system, transportation is still easier and more available than in rural areas.

Urban Stresses

While there are certainly many positive reasons why people are drawn or pushed to urban areas, there are also many negative attributes to urbanization. Two important problems related to urbanization, whether in the U.S. or anywhere else in the world, are social stresses, those with a direct impact on people; and environmental stresses those with an impact on the physical environment.

Social Stresses

In both developed and less developed countries, the increased number of people, and the resulting increase in the density of those people, often bring on incredible despair. The crowding of more people into the same limited space, the poverty brought on by a too-ample supply of people willing to compete for under-paying jobs, and the fear brought on by living in areas saturated by crime are some of the social stresses to which people in urban areas are subjected.

Poverty

Many poor people from rural areas move to the city to escape poverty. Unfortunately, jobs in the cities are not as plentiful, nor is the quality of jobs nearly as high, as many urban immigrants believe. As a result, many urban immigrants are unemployed, and many others are significantly under-employed. They are transformed not from rural poverty to urban affluence, but simply from rural to urban poverty.

As a result of this poverty, the urban poor often live in abject squalor. In many cities in both developed and less developed countries, thousands or even millions of people are literally homeless, living on the streets, in parks or anywhere else they are allowed. Thousands of other people live in cardboard shanties. In the U.S., the most affluent country in the world, it is common to step around homeless people sleeping on the sidewalk within sight of the White House and the Capitol. No city in the U.S. is without its ghetto, or barrio, or slum; no city is immune. Another consequence of poverty, including urban poverty, is malnutrition. Many of the urban poor, like their rural counterparts, simply cannot afford adequate food for their families, even though a grocery market might be located in the immediate vicinity.

Compounding the effects of malnutrition, poverty prevents the poor from receiving adequate health care. They are unlikely to have even basic sanitary facilities or sources of clean water, and insect-and water-borne diseases run rampant. As a result, their quality of life is diminished, and their anticipated life span is significantly shortened.

The situation is even worse in the urban areas of less developed countries. It is estimated that in Calcutta, India, more than 70% of its approximately 12 million people live below the poverty level, and more than a half a million are homeless. The plight of the poor in Mexico City was shown to the world a few years ago when a massive earthquake killed tens of thousands of people, many of whom lived in cardboard or tin shelters.

Crowding

Crowding is a significant source of stress in many urban areas. Space is often expensive and hard to find. The result is that the majority of people is forced to live in less than optimum space and conditions. Crowding has been blamed for many social problems that seem to be the most severe in urban areas. These problems include a variety of physical, emotional and social ills, one of which is an unusually high prenatal death rate. In addition, social scientists believe that crowding is one of the factors producing the drug and alcohol abuse that seems to run rampant in crowded urban areas.

Crime Rate

Crime rates are higher in poor urban areas than in affluent urban, suburban, or rural areas. There is certainly a cause and effect relationship between poverty and crime. Teen-age ethnic gangs roam the streets with apparent impunity, establishing the local 'laws.' Anyone who tries to improve the situation is subject to killing or maiming. In recent years, senseless drive-by shootings have significantly increased. Robberies are often the result of poverty, but too often the result of the need for drugs and/or alcohol. The value of a single life is not thought to be very high.

Due in part to the crime rate in poor urban areas, a growing sense of despair is developing among the young people who live there. It is not uncommon to hear a pre-teen or young teen talk objectively of his or her likelihood of being killed before adulthood. This sense of despair leads to an attitude of "get what you can, while you can, and don't worry about the consequences."

STRESSES ON THE PHYSICAL ENVIRONMENT

In both developed and less developed countries, the increased number, and increased density of people in urban areas bring on severe stresses to the physical environment. The stresses to which the urban physical environment are subjected include effects on Air Quality, on Water Quality, on Water Resources, and on Waste Disposal.

Air Quality

Nearly all urban areas have a negative effect on air quality. The problem of air pollution is discussed in detail elsewhere. Only those problems specific to urban areas or problems exacerbated by urbanization will be discussed here.

In most urban areas, automobiles are the single most important contributor of urban air pollution. In automobile-congested cities like Los Angeles, Sao Paulo, Mexico City and Bangkok, automobiles are responsible for as much as 85% of all air pollution. Incompletely burned hydrocarbons from gasoline and diesel fuel react with nitrogen oxides in the presence of sunlight to form a "brown air," or photochemical smog. Cities with effective mass transit systems, like Washington, DC and San Francisco, are less often plagued with this type of air pollution than cities more dependent on automobiles.

In the recent past, many cities that relied on the burning of coal and heavy oil for the production of energy were plagued by another form of smog. "Gray air" smog, produced by suspended particles, is still a major contributor to air pollution in many urban areas in less developed countries, including China and many eastern European countries. In developed countries where better pollution abatement procedures are followed, urban areas now have significantly less of a problem with gray air smog.

Photochemical smog is a significant problem in all urban areas, but some large cities have particularly harsh smog problems. The local climate and topography of some urban areas combine to inhibit the dispersion of photochemical smog by natural means, such as wind. Temperature or "thermal" inversions occur when weather conditions trap a layer of colder, more dense air beneath a layer of warmer, less dense air, effectively blocking the upward dissemination of pollutants. Los Angeles is severely affected by this phenomenon, and in some cases, it raises the concentration of pollutants to harmful, even lethal, levels.

Water Quality

High concentrations of people in urban areas can significantly affect the quality of the water supply. The problem of water pollution is discussed in detail elsewhere. Only those problems specific to urban areas or problems exacerbated by urbanization will be discussed here.

Urban areas produce a great deal of waste, and a high percentage of this waste is placed in sewage. A city's ability to clean and dispose of sewage is made more difficult by the great volume of water that must be handled. In most cities, the drainage from streets and buildings is combined with other types of sewage. Since most city areas are paved or covered by roofs, the majority of the rainwater that falls in a city eventually finds its way to the city's sewage treatment facilities. During times of heavy rain, these facilities are often overwhelmed and must release water without sufficient treatment. Not only does this practice lead to pollution in local lakes and streams, it can also flood areas downstream.

Water Resources

Urbanization places tremendous stresses on local water resources. Water resources for urban areas come from either surface water or ground water.

Many urban areas derive their water from adjacent rivers or lakes. Some of the surface water is diverted to local water treatment plants for purification before being used. In these cases problems arise only if the demand for water by the population being served by the treatment facility exceeds either the amount of surface water available, or exceeds the treatment capability of the water facility. It is incumbent upon urban planners to keep in mind the limits of local water resources. If these limits are exceeded, urban areas must find additional water from other resources.

Urban areas that derive their water from ground water sources must always be mindful of the possibility of aquifer drawdown, which results from excessive ground water extraction. Problems arise when water demands on the aquifer exceed the rate at which the aquifer is naturally recharged. These problems are often exacerbated in coastal areas, where aquifer drawdown can lead to salt water intrusion. If the aquifer is below an urban area, the city must make certain that the water recharging the aquifer does not become polluted before or after it soaks in the ground.

Waste Disposal

Rapid population growth in urban areas also places great stresses on municipal waste disposal systems. Sewage treatment systems and municipal solid waste disposal systems, whether landfills or incinerators, have limited capacity. The problem is especially acute in areas where the soil and physiographic conditions limit the number of waste disposal sites that can be established, and thereby limit the total volume of material that can be disposed.

Emigration from Urban Areas

Over the past 40-50 years, particularly in the United States, certain populations have been moving out of urban areas. During this time there has been increasing migration from large central cities to suburbs and adjoining smaller cities. The reasons given for this emigration include:

- Escape from Urban Stresses
- Improved Infrastructure
- Establishment of New Employment Opportunities
- Perception of Better Educational Opportunities

Unfortunately, the emigration from the central city to suburban or adjacent smaller communities has a negative monetary effect on the central cities. When businesses and the more affluent segment of society move away from the central city, so do the tax revenues they generate. Budget crunches have a severe effect on the abilities of central cities to address the reasons for this emigration, and attempt to reverse it.

Escape from Urban Stresses

The single most important factor in the emigration of people from central cities to suburbs or smaller adjacent communities is to escape urban stress. For many years, the emigration to suburbs resulted in racial and ethnic resegregation. Today, this emigration appears to produce less racial and ethnic segregation, but more socio-economic segregation. People of all races who are part of the lower socio-economic class tend to remain in the central city, and people in the middle and upper-middle socio-economic classes, whatever their race, tend to move to the suburbs. When families become economically able to do so, they move away from the crowding, poverty, crime, pollution and waste disposal problems associated with many central cities.

Improved Infrastructure

One significant reason for emigration from central cities to suburbs is that larger and more efficient streets and highways now connect the central city with outlying areas. It is easier now to get into the city for business, education or pleasure, and conversely, to return to one's living area. Some cities, most notably Washington, DC, New York City, Atlanta and San Francisco, have mass transit systems that extend into surrounding suburban areas, but such efficient systems are the exception, not the rule, among large cities. Most of the movement of commuters between central cities and suburbs takes place in automobiles-and these are most frequently occupied by only one person. Car-pooling is encouraged, but rarely required or rewarded.

Establishment of New Employment Opportunities

An important factor in the emigration from central cities to suburbs has been the location of new businesses in, and the relocation of established businesses into, the areas surrounding cities. This is especially true for smaller businesses and light industry, since they do not need to be in a city center. In most developed countries, efficient and dependable communication and transportation facilities are readily available nearly everywhere, making it possible to do business wherever one wishes. Many business managers and CEOs feel that the less crowded, less hectic, and more aesthetically pleasing suburban or small city environment results in greater worker productivity. As businesses move, workers follow and populations grow in the suburbs and in smaller cities.

Perception of Better Educational Opportunities

A large factor driving emigration from central cities is the perception that suburban schools are better than their central city counterparts. While central city schools are not irrevocably doomed to decay and sub-standard education, there is no doubt that these schools are affected by the declining tax base that accompanies the loss of businesses and the more affluent portions of society. As available funds decline, the physical and educational quality of the schools will decline as well. What can develop is a classic case of 'the rich getting richer, and the poor getting poorer,' and the educational gap begins to widen between the 'haves' and the 'have-nots'.

Coastal Land Use

The world's coastlines are being subjected to increasing pressure as populations grow, and peoples' needs for transportation, food and recreation expand. Coasts and coastal areas serve a great variety of functions, and each must compete with the others for available space and clean water. In the United States, over 50% of the population lives within 50 miles of a coastline and more of our population is moving to the coast every year. Effective preservation of these valuable lands will have to be based on a knowledge of Coasts and Coastal Processes; an understanding of Coastal Legislation, and an appreciation of the pressures being exerted by a growing Population.

Coastal Processes

Most changes along the coast are produced by waves. In some places it is possible to see waves accomplish a significant amount of change in only a few hours. The basic process is fairly simple. Most waves are produced by storms out at sea. As the waves approach the coast, they pick up and move coastal materials. The size and orientation of the waves determines what they can do. Small waves tend to move sand and other materials onto a beach, large waves move materials off of a beach, and waves that meet the coast at an angle move materials along the beach. The amount of change that waves can bring about depends in part on the type of coastline they encounter. South Carolina's coastline is low and flat. This means that it generally receives small to moderate waves, and that changes in sea level can have a dramatic effect. Like most coastlines, South Carolina's should maintain a dynamic equilibrium, or a balance between all of these changes. As waves move sand away from a beach, there should be a matching supply of sand being added to the beach. As rising sea level moves the beach landward, the sand should be able to move with it and there should be room for the new beach to form. As people develop coastlines to create ports, resorts, and marinas, we must make sure that we do not obstruct the natural processes that maintain the beach. If we do not understand how the coastline works, it is likely that we will destroy it.

Origin of Waves

When energy from the wind is transferred to the water, waves are produced, and the waves then carry that energy as they travel for hundreds or thousands of miles. This is possible because it is energy that is moving, not water. It is easy to demonstrate this process by floating a ping-pong ball on waves in an area free of wind. The ball will move up and over each wave, and return nearly to its original location. Each wave moves the ball around in a circle, and since the ball is in contact with the water, the water must be moving in the same way. Waves represent energy passing along the surface of the water in much the same way that "waves" of energy can pass through a rug or a rope as one end is shaken up and down. In the case of waves in deep water, this circular motion is confined largely to the upper layers. Observation indicates that the waves' motion decreases with depth and nearly dies out at a distance below the waves equal to one half of their length. Thus, if waves are found to be about 40 feet long (the distance from one crest to the next), we would know that they are disturbing the water to a depth of about 20 feet. This point is known as the "wave base." Most waves in the ocean, even the very large ones, have a base no more than 40 or 50 feet deep. Since most of the oceans are several thousand feet deep, a 50-foot base is barely disturbing the surface, although such a wave would seem huge to us. This surface motion means that waves can pass fairly easily from one place to the next without a significant loss of energy. Thus the world's oceans become great collectors of energy that is steadily transported toward the coastlines. When the "waves of energy" reach the coast, they can accomplish a great deal of work by expending this energy over a small area in a short period of time.

Size and Orientation of Waves

As waves approach shallow water, they eventually reach the point at which the wave base is equal to the water's depth. As the bottom begins to interfere with the circular motion of the waves, it uses up some of their energy and begins to slow them down. When the leading waves slow down, the following ones push in behind, and the result is that the waves are squeezed together and pushed upward to become larger. As the waves become taller, they become more unstable until they fall forward and "break" on the beach. Each breaking wave can pick up a sizable load of sand, shells, or whatever is available and move it forcefully up the beach. If closely watched, this beach material will be seen to move in a number of directions:

- The most obvious motion is up onto the beach and back off again. As waves pick up sand and move up onto a beach, some of their water soaks into the beach. Since small waves contain less water, they lose a greater percentage of their volume and thus a significant amount of their ability to move sand. Larger waves hold greater volumes of water and thus lose a smaller percentage into the sand. This means that large waves can hold their loads in suspension long enough to carry the sand not only onto the beach, but also back out again. The result is that large waves tend to transport material off the beaches, while smaller waves with less transporting ability tend to build beaches up. This may be seen on many beaches that become smaller during the winter when waves are large, and grow back in summer when waves are smaller.

- The second motion is along the beach. The careful observer will note that most waves approach the shore at some angle, not straight-on. This gives the waves the ability to carry material along the beach as well as on and off of it. This means that the sand picked up by a breaking wave will eventually be deposited some distance down the beach. As each wave repeats this process, a sizable amount of sand is moved along the beach face. This is one component of a process known as "longshore transport" that carries many millions of tons of sand and other beach material along coastlines every year.
- The third motion is the most difficult to see, but easier to experience. While playing in the surf many people realize that they are being slowly carried down the beach. This second component of longshore transport is a current that can occur in the surf zone. As waves approach the beach at an angle and break, a large amount of water is dumped into a "trough" between the beach face and adjacent rows of sand bars. The breaking waves add water to the trough, but there is no easy way for the water to escape. Seeking an outlet, the water flows along the beach carrying sand with it until it finds a gap between bars. When a gap is found the water escapes to the ocean in a strong current known as a rip. Though short and narrow, these currents account for what is often called "undertow".

Changes in Sea Level

Variations in sea level can have a dramatic effect on a coastline. Some changes in the ocean's level are small and frequent, some are very slow and gradual, and still others are rare and precipitous. The most obvious changes are the regular rise and fall of the tide. These changes are generally well known, and range from inches in some areas to 40 or 50 feet in others. Along South Carolina's open ocean beaches, the difference between high and low tide is usually between five and seven feet. Coastal residents are well aware of tidal fluctuation and plan for it, but their plans often do not include the slower changes in sea level that occur over decades and centuries. For reasons that are not well understood, the earth's climate has passed through cycles of warming and cooling for eons. When the climate begins to cool, polar ice cover increases and fresh water that would have run into the ocean is tied up as ice. The result is a drop in sea level. During the last ice age, sea level fell over 300 feet. During warming times, polar ice melts and sea level rises. Over the past 10,000 years, the earth's temperature has been increasing and the sea has been rising. During the last century the rate of warming has increased and the rise in sea level is accelerating. Evidence for this rise can be seen in ancient structures that are being flooded and coastal areas that are being invaded by the sea. There is little that humans can do to slow this rise, and thus our coastal building plans must take it into account.

Perhaps the most dangerous rises in sea level are the rapid and often unpredictable "storm surges" that accompany hurricanes and other severe storms. Storms can bring about a dramatic rise in local sea level in two ways. First of all, most large storms are associated with an area of low atmospheric pressure. As atmospheric pressure decreases, the sea bulges up just the way a balloon or a sealed bag would "bulge" as it is taken up a mountain or up in an airplane. Second of all, the effect of this bulge can be made more serious by a strong on-shore wind that "piles" the water up against the coast. The resulting surge of water can produce a "wall of water" ten to twenty feet high that can do a great deal of damage as it moves onto the coast.

TYPES OF COASTLINES

Coasts look very different from one place to the next. Compare the rocky coast of Maine or California with the sandy coasts of the Carolinas or the muddy coast of Louisiana. Some of this variation is due to the material that forms each coast, and some is due to the amount of energy the coastal area receives from waves. Most coasts have some sort of beach where these materials are temporarily deposited. At one end of the spectrum is the very irregular muddy "beach". In areas like this, the ocean's waves are so small and infrequent that they cannot bring about any significant erosion or transportation of coastal materials: In this very low energy environment, the beach is made up of whatever material is deposited by processes on land. Saltmarshes are common along this type of coast.

Further along the spectrum is the type of coastline that is exposed to waves of moderate energy. In this case, the processes of longshore transport work well most of the time. The very lightest materials (particles of silt and clay) are easily carried away, while the heavier sands and light gravels are spread out along the coast to form smooth, straight, sandy beaches.

At the high energy end of the spectrum, the waves are large and able to cut away and transport almost any material. The result is usually a steep, rocky beach that lies at the foot of a cliff. This type of beach changes frequently as the waves cut away at the base of the cliff and advance the beach landward. Only the heaviest materials are left on the beach, while the rest are soon washed away.

The Beach

The word "beach" is familiar to most people, but it has a specific geologic definition. A beach is the area of a coastline that is affected by waves. It extends from the highest point on land that is reached by storm waves, out into the ocean to the point where the waves' bases first contact the bottom. (For convenience, this point is said to be a depth of 30 feet.) Beaches can be composed of any material that can be moved by waves. Some are made of boulders, some are formed from shells and shell fragments, but most beaches have some sort of sand.

Sand originates on land from the physical and chemical weathering of rock. Generally it is transported to the coast by streams and is distributed along the beach by the action of waves. Sand is stored temporarily in dunes and sand bars, but over a long period of time, sand is in motion on and off the beach and along the beach. Most of this motion is due to the action of waves, and the amount of work waves can do varies greatly from place to place.

Salt Marshes

Low energy coasts in protected areas such as estuaries and behind barrier islands collect fine grained sediments (muds and silt) that support plant growth. These areas benefit from their proximity to land and their connection to the ocean. Land areas are a source of nutrients, and the ocean provides stability. The result is that Saltmarshes contain some of the most productive land on earth. An acre of open ocean might be able to produce half a ton of new organic matter each year, a cultivated wheat field might be able to produce about two tons; but saltmarshes can produce nearly ten tons of organic matter per year. This tremendous productivity combined with the stable conditions maintained by the adjacent ocean make saltmarshes ideal nurseries for young sea life. Along the southeastern coast of the U.S., nearly 90% of the commercial fish spend a portion of their lifetime in saltmarshes. Saltmarshes are also excellent filters that can remove natural and human-produced pollutants from coastal waters.

As useful and important as they are, saltmarshes are under pressure in many areas of the country. Marshes lie in the path of coastal resort development, marshes receive much of the pollution sent down rivers, and marshes are often caught between rising sea level and armored coasts. Until recently marshes were seen as useless flat land that should be filled or drained.

COASTAL LEGISLATION

As states such as South Carolina became increasingly aware of the value of their coastal areas, they realized that many of these areas could be lost if they are not protected by strong and realistic legislation. The legislation is needed to protect public lands from damage by private actions, but at the same time the laws needed to be sensitive to the rights of private land owners. Many people do not believe that South Carolina's legislation has achieved the proper balance between these two needs, but most people believe that action is important even if it is not perfect. South Carolina's action in this area is the South Carolina Coastal Zone Management Act of 1977 and the Beachfront Management Act of 1988. The basic purpose of this legislation is to remove any type of construction from the most active part of the beach. If the beach's natural processes are to work, we must not do anything that will interfere with the natural flow of sand and other beach materials. These acts attempt to accomplish this with three primary provisions:

- *Base Line:* A base line has been established along the entire coast. This line is the "Primary Dune Line", or a line marking the dunes that are closest to the ocean. This line indicates where the dunes are, or where they would be if they had not been removed. The BFA does not permit any type of construction between the base line and the ocean. Landowners may not take any action to armor the coast or affect the flow of the sand in that area, and no buildings of any sort can be placed there. Existing structures in this area do not have to be removed, but if they are damaged, they cannot be repaired or replaced.
- *Set Back:* Any new construction must be "set back" behind (landward of) the base line. The amount of set back depends on the rate at which the coastline is changing. The set back must be 20 feet, or 40 times the yearly rate of erosion whichever is greater. For example, if the coast in one area is eroding at one foot per year, any new construction must be set back 40 feet behind the base line. If no detectable erosion is occurring, the construction must be at least 20 feet behind the base line. Existing structures within the set back zone cannot be over 5,000 square feet in size, and if they are more than 50% destroyed, they cannot be repaired or replaced.
- *Sea Walls:* Existing sea walls can be repaired or replaced as long as the new structure is tilted backward (away from the ocean) so that the energy of the waves is not directed downward toward the beach, but upward away from the beach. Without this type of design, a sea wall will destroy the beach in front of it.

POPULATION

The world's total human population is growing, and this growth is seen on most scales from whole counties to cities within countries. Obviously there are exceptions. There are places with stable or shrinking populations, and there are places whose populations are growing more rapidly than others The world's coastlines are one of these exceptions. For one thing, a majority of the world's cities are on coastlines. In the U.S., for example, twenty of our largest twenty five cities are on a salt or fresh water coast. Since the population of most cities is growing disproportionately rapidly, this adds large numbers of people to some coastal areas. With very few exceptions, these cities were established along a coast because of the easy transportation provided by the adjacent water. Non-urban coasts are also growing due to their value for food production and recreation. These various uses of the coast compete with each other for space, clean water, and access to the ocean. Each is going to have to accept limitations on its activities if all are going to survive and prosper.

Water Transportation

Throughout most of human history, water transportation was the only way to move large loads over long distances. Today there are other choices, but water transportation is still the least expensive. Many tons of bulk material such as ore, coal and grain are moved by water each year. Water transportation on a large scale, requires a large amount of construction on and off shore. Channels and port facilities have to be built and maintained, and land-based storage and transportation needs to be readily accessible. Each of these has an effect on the coastal environment. When channels are built and maintained, the dredge spoil (sediment that is removed by dredges) must be dumped somewhere. If dumped in the water, it can cover bottom dwelling organisms and muddy the water. If the spoil is dumped on land, it can cover valuable land or alter drainage patterns. Port facilities require construction at the water's edge, and unobstructed access to open water. Each of these can serious impair the natural movement of sand and other materials along the coast. In South Carolina, for example, the maintenance of port facilities at Charleston severely limits the natural flow of sand southward along the coast. This limit contributes to the erosion of Morris Island and Folly Beach. Finally, the construction of warehouses and truck and railway terminals along the coast uses valuable land and seriously alters the shoreline and natural shoreline processes.

Food

Coastal people have always relied on the adjacent water for food. For many people there may be no other available sources of food, and for many others the ocean is being touted as the source of food for the future. Environmentalists foresee some significant problems with this on several levels. First of all, the open ocean is not an environmentally efficient source of food since most of its edible inhabitants feed at a high trophic level. This means that there are many steps between the marine plants at the bottom of the food chain and the edible fish near the top. Shell fish and marine plants are closer to the bottom of the food chain. However, the second significant problem is pollution. Humans have dumped waste into the ocean for centuries and continue to do so. Though this has led to significant levels of pollution in all parts of the ocean, the problem is and always has been greatest along coats where waters and shallow and sources of pollution are nearby. These are the waters from which most shell fish is harvested. Many people believe that the solution is marine aquaculture whereby large portions of the coast would be closed off, protected, and used to grow a variety of shell fish and crustaceans. The problem, of course, is that there that are many other demands for the same areas. The most significant demand in many areas is for housing and recreation as increasing numbers of people wish to live or vacation at the coast.

Recreation Purpose

The invention of the American summer beach resort has brought millions of people to the coast to spend a day, a week or even a year. If the people are coming, there must be a places for them to stay, they must have some place to eat, and they must have things to do with their time. To be the most successful, all of this should be within view of the water. The result has been the construction of hotels, cabins, restaurants, marinas, amusement parks, and shopping centers on land that is temporary and transient. In many cases, the construction has included the removal of dunes and ancient beaches so that the ocean is more visible and readily accessible. Coastal legislation and common sense have limited some of the earlier abuses, but an unfortunate vicious circle still remains. If the land is naturally temporary and transient, while simultaneously being very valuable, the landowners feel obliged to take any measures possible to prevent change. These measures include items such as: sea walls built along the back of the beach to keep the waves from reaching beyond that point, groins built perpendicular to the beach to trap and hold sand as it moves along the shore, and jetties built at the mouths of inlets to keep sand from flowing in. All of these devices interfere with the natural movement of sand along the coast, and in the end contribute to the destruction of the coastline they were designed to protect. Experts agree that successful preservation of valuable and scenic coastline involves retreat from geologically active areas. Only if the natural coastal processes are allowed to operate unimpeded can the coast be preserved.

3

Pollution

AIR POLLUTION

Air is a vast renewable resource used by all living things. Unfortunately, the air has also long been a dumping site for tons of unwanted material. Since the air is usually mixed rapidly and thoroughly, whatever is put into it is soon distributed over a wide area. To a certain extent this has always been a part of the "solution"; to disperse pollutants broadly and quickly so that they are no longer a problem.

More significantly, however, the fact that the air easily carries pollutants from one area to the next means that one person's "solution" becomes everyone's problem. Air is a rather stable mixture of nitrogen, oxygen, and water vapor, with much smaller amounts of carbon dioxide and several inert gases such as argon and helium. Any changes to this mixture could be considered pollution. Most types of air pollution can divided into two major categories: pollution by the addition of particles, Particulate Pollution, and pollution by the addition of gases, Gas Pollution.

Particulate Pollution

All but the very cleanest air contains small particles of liquid and solid material. By definition, air is composed solely of gases, and thus these particles can be considered pollutants. Some particles have a natural origin, and other are produced by humans. Whatever their origin, particles in the atmosphere can cause problems. The most obvious problem is appearance. Soot and dust in the air look bad, and the particles eventually settle out of the air and accumulate on leaves, automobiles, and window sills. For many people, particulate pollution is also a significant health problem. Pollen can produce allergic reactions, and prolonged exposure to coal dust and finely powdered rock can produce lung diseases such as black lung and silicosis. On a much larger scale, particles in the atmosphere can lead to the atmospheric condition for which the word "smog" was first coined almost 100 years ago.

Smog

As most people know, the word "smog" comes from a combination of the words "smoke" and "fog". What might be less familiar is the origin of this term about a century ago in London. At that time, soft coal was the city's primary fuel in factories and homes. Burning soft coal can make a great deal of smoke. London's climate, particularly in the winter, is cold and damp. When moisture-laden air is cooled (as it might be on a bathroom window) the moisture condenses on any available surface. That surface might be a window pane, a water pipe or a cold drink can. Particles suspended in the air also provide surfaces for condensation, and when moisture condenses on particles, they get larger, creating a fog. Thus, smoke can produce fog. Smoke was the primary cause of the famous London fogs, and when the burning of soft coal was curtailed, the fogs began to disappear.

Natural Causes: Many particles are placed into the atmosphere by natural processes which are unrelated to human activity. Pollen enters the air from flowers and salt crystals form when ocean spray dries. Dust storms and volcanic eruptions put dust into the atmosphere, and forest fires create smoke which is composed predominantly of small particles of ash and minute pieces of unburned carbon.

Human Causes: Smoke contains most of the particles which enter the atmosphere as the result of human actions. Visible smoke can be composed of a variety of particles such as soot, unburned fuel and minute pieces of ash. For example, coal-fired electric power stations usually grind coal into a fine power before burning it. When the powdered coal burns, it makes minute particles of ash which might be carried up the smoke stack and create a gray smoke. Automobiles which are not operating properly often discharge small particles of oil which make a blue smoke, or small amounts of unburned gasoline making a black smoke. Diesel engines often discharge a black smoke composed of particles of soot.

Solutions: People have developed a variety of means to deal with air pollution. Since particulate pollution was the first and most obvious problem, its solutions came first. There are four common ways to combat particulate pollution:

(1) *Bag Filters:* Much like giant vacuum cleaner bags, bag filters are suspended inside chimneys and smoke stacks. The smoke or other waste gas is forced to pass through the bag, and the majority of the associated particles are trapped. Some industries collect the particles in truck loads and sell them for use in construction materials such as concrete blocks.

(2) *Cyclones:* As the name suggests, a cyclone is a mechanism to spin waste gases, forcing the denser particles to the outside and into a collecting devise. Textile mills often use this system to remove lint from the air in manufacturing areas.

(3) *Scrubbers:* These devises use a mist of water to "wash" pollution out of the air. As the polluted air passes through the mist, water droplets stick to the particles and make them heavy enough to fall out of the air. The air is cleaned in much the same way that a rain storm can clean pollen out of the atmosphere. Scrubbers are one of the systems which are effective against both types of pollution. They can also wash some types of gas pollution out of the air. For example, sulfur dioxide (SO_2) will dissolve in water creating an acid rain inside the scrubber rather than outside in a forest.

(4) *Electro-static Precipitators:* The electro-static precipitator uses static electricity to give the polluting particles a strong electric charge. After receiving their charge the particles pass through a screen with the opposite electric change, they are attracted to the screen and attach themselves to it. The screen must be washed periodically to remove the particles.

GAS POLLUTION

Any change in the atmosphere's natural mixture of gases has the potential to be significant air pollution. Most gases are invisible, and thus this type of pollution is less obvious. Since it is out-of-sight, it was for many years out of mind. Today, however, gas pollution is a familiar and important problem. Exotic gases can enter the atmosphere from natural as well as human sources. Once in the atmosphere, the gases may cause problems alone or they may form troublesome compounds by combining with particles or other gases.

Natural Sources

Several important air pollutants have natural sources. One is methane (CH_4). Methane is produced by the bacteria that break down organic matter in the absence of oxygen. This process can occur in swamps, on the sea floor, and inside the bodies of animals. The methane which escapes to the atmosphere adds to the "Greenhouse Effect". Natural fires produce carbon dioxide (CO_2) which is also a green house gas, and volcanoes produce a great deal of Chlorine gas which may have an effect on the Ozone Layer.

Human Sources

In industry, domestic heating and many varieties of transportation, humans produce gases that are released into the atmosphere. In most cases these gases affect the Lower Atmosphere into which they are released, but in some cases they can affect the Upper Atmosphere.

Lower Atmosphere

The most prominent air polluters are: Sulfur Dioxide (SO_2), Nitrogen Oxides (NOx), Carbon Monoxide (CO), Carbon Dioxide (CO_2) and Chlorofluorocarbons (CFCs).

Sulfur Dioxide

Sulfur dioxide (SO_2) is a colorless gas that is produced when the element sulfur is burned. This can happen intentionally during the purification of ores which contain sulfur. In a process known as "smelting", the ore is heated to a high temperature and the sulfur is burned off. Sulfur is burned unintentionally when coal or other fuels containing a large amount of sulfur are used by industries and public utility companies. Whatever its source, once the sulfur enters the atmosphere it combines with water to form sulfuric acid (H_2SO_4) and becomes an important component of acid rain.

Nitrogen Oxides

Nitrogen oxides are produced when a mixture of nitrogen and oxygen is subjected to high temperatures and high pressures. These conditions occur continuously inside the internal combustion engines that operate automobiles, trucks, trains and small airplanes. These two primary atmospheric gases are drawn into the engines, some of the oxygen is used up as the fuel burns, but the remainder is heated and compressed inside each cylinder as it fires. The resulting nitrogen oxides are responsible for three important air pollution problems.

Effects of Lower Atmosphere Gas Pollution

There are several biologic and/or non-biologic effects of lower atmosphere gas pollution. Included among these are the following:

(1) Acid precipitation has its source in the combustion of fossil fuels. When both sulfur and nitrous oxides are emitted into the atmosphere, both can react with water vapor to produce sulfuric acid (H_2SO_4) and nitric acid (HNO_3), respectively. Both have both biologic and non-biologic effects. Biologic effects include acidification of soils, plants and surface and ground water. Acidification can stunt or even kill the growth of plants by acting on the plant's leaves or by ingestion through the root system. An excellent example of the detrimental effects can be seen in the destruction of entire stands of fir and spruce trees in the Blue Ridge and Great Smoky Mountains of North Carolina and Tennessee. Acidification of soils can result in the alteration of soil chemistry by leaching acid-soluble chemicals and effectively increasing the levels of certain other non-acid-soluble chemicals. Runoff of acidic soils in upstate New York in the 1950's led to the acidification of Lake Erie and the near death of that

Great Lake. In recent years, Lake Erie has been brought back to life as a result of practices designed to reduce soil acidification. Acid precipitation has even been linked to human respiratory diseases such as asthma and bronchitis.

Non-biologic effects of acid precipitation include the dissolution and destruction of many statues and buildings, especially those made of limestone and marble, and the "pitting" of many metal and painted surfaces.

(2) *Ozone:* In the presence of sunlight, one of the atoms of oxygen is often driven off of nitrogen dioxide (NO_2) and is free to combine with an oxygen molecule (O_2) to form ozone (O_3). Ozone is beneficial to humans when in the upper atmosphere, but in the lower atmosphere ozone is an irritant making throats sore, eyes red, and noses run.

(3) *Photochemical Smog:* In the presence of sunlight, nitrogen oxides can combine with unburned hydrocarbons (gasoline and oil) to form chemicals such as formaldehyde and peroxyacetyl nitrate (PAN) which are some the primary components of the brownish haze which often forms over Los Angeles and other sun belt cities during the summer time. Since this new type of "smog" is formed by chemical reactions that are driven by sun light, it is known as Photochemical Smog to distinguish it from the earlier type of smog.

Carbon Monoxide

Carbon monoxide is produced when a compound containing carbon is burned at a very high temperature, or burned without a sufficient supply of oxygen. This occurs in internal combustion engines (cars, trucks and buses), in trash piles, and in many types of domestic heaters. Carbon monoxide is poisonous because it is absorbed by the blood in place of oxygen. If we breath too much carbon monoxide, we can die from a lack of oxygen.

Carbon Dioxide

Carbon dioxide (CO_2) is produced when any material containing carbon is burned. Whether the material is wood, oil, garbage, paper or natural gas, as it burns oxygen combines with carbon to make heat and give off CO_2. Carbon dioxide is also produced by all living things when they obtain energy from carbon-based food by combining it with oxygen and giving off CO_2 as a waste product. Carbon dioxide is a natural, though quite small, component of the atmosphere, and for this reason, it was not considered an air pollutant for many years.

Environmentalists became concerned about carbon dioxide when it became apparent that the atmosphere's average temperature was increasing worldwide. Since carbon dioxide absorbs infra-red radiation (heat), this increase in temperature could be explained by an increase in the amount of CO_2 in the atmosphere. This situation, known as the "greenhouse" effect, is the source of increasing concern. First of all, because a number of gases which we have released into the atmosphere have the same effect as carbon dioxide in that they absorb infra-red radiation. Called "greenhouse Gases" they include nitrogen oxides, chlorofluorocarbons, methane (natural gas), water and carbon dioxide. Second of all, a general warming of the atmosphere can lead to a great variety of environmental problems.

Global Change

The term "Global Change" refers to all of the many ways that the earth appears to have changed or be changing during the time that humans are here. Some of the changes are the direct result of human activity, some clearly are not, and some are not well enough understood to attribute to any cause. This section will look at the global change issue, some of its causes, some of its effects on the Earth's sub-systems, and some of the actions that are being taken or could be taken to address the issue. The global change issues to be examined include:

Global Warming

One can hardly escape the reports in the media about global warming and the greenhouse effect. To understand the concept of global warming, it is necessary to remember that throughout geologic history, the global climate has warmed and cooled several times, resulting in Ice Ages when the global climate was significantly colder than present, and Interglacial Stages, when the global climate was significantly warmer than present. In fact, during the last 2 million years, there is now evidence of at least 20 glacial ages.

To understand the effect that human activities have had on global climate, we must understand where we are in the glacial/interglacial time-frame. The consensus among earth scientists is that we are in a global warming period that was not initiated by human activity. Because we are concerned with the possibility that human activity might further warm the global climate, we must attempt to separate the human-induced and the non-human-induced components of global warming.

Upper Atmosphere Ozone Layer

The stratosphere is that layer of the atmosphere that extends from 17 km to 48 km above sea level. Within the lower parts of the stratosphere there exists a "layer" rich in ozone gas. Ozone, whose chemical formula is O_3, is important because it absorbs ultraviolet radiation from the sun, and thus limits the amount of that radiation which reaches the earth's surface. Recent atmospheric studies show that the amount of ozone in the upper atmosphere varies from one location to another, and from one time to the next.

This means that the amount of ultraviolet radiation reaching the earth also varies. This is important because ultraviolet radiation is known to be harmful to many living things, including humans. Any reduction in the amount of ozone in the upper atmosphere can have detrimental effects. It has been well documented in the past few decades that there is a roughly circular area in the ozone layer where the ozone concentration is dangerously low.

This hole in the ozone layer is not constant in size or location, but it appears to be most problematic in the Southern Hemisphere, over Antarctica, Australia and New Zealand. Unlike many environmental issues, where we do not know for sure the causes and effects of the problem, in this case we do. One of the primary causes for the reduction of ozone in the upper atmosphere is an increase in a series of man-made chlorine-and fluorine-compounds known as chlorofluorocarbons.

Greenhouse Effect

The Earth's atmosphere makes it possible for life to exist. If there were no atmosphere, the surface temperature would be about 32 degrees colder than it is now. The amount of radiation reaching and warming the Earth must balance the amount leaving and cooling the Earth. This delicate balance is known as the radiation budget. The atmospheric gases allow the sun's short wave length radiation to pass through to the Earth's surface. The surface warms and re-radiates long wave length, infrared (heat) radiation to the atmosphere. This is what makes the Earth's temperature "just right" for life to exist.

Gases in the atmosphere allow incoming, short wave length radiation to pass through the atmosphere to the solid Earth and hydrosphere. Certain of these gases, however, do not allow the outgoing longer wave length radiation to re-radiate through them, thereby "trapping" the longer wave length infrared radiation. This is analogous to how a green house works. In a green house, the glass or plastic roof and walls allow incoming, short wave length radiation to pass through, but do not allow the outgoing, longer wave length infrared radiation to pass back out, raising the temperature in the green house. This gives rise to the term Greenhouse Effect for the process by which outgoing, longer wave length infrared radiation is trapped in the atmosphere, raising the atmospheric temperature.

The gases in the atmosphere that absorb the most infrared radiation are water vapor and carbon dioxide, but methane from plant decay and animal digestion, nitrous oxides from fertilizers, and man-made gases called chlorofluorocarbons or CFCs also absorb heat. Together these are known as greenhouse gases, or those gases which produce the greenhouse effect. We need carbon dioxide and the other gases in our atmosphere to help hold heat on the Earth's surface. However, increases in their concentrations due to the extensive use of fossil fuels as our primary energy sources can lead to higher atmospheric temperatures.

It is this increase in atmospheric temperature that is the principal component of human-induced global warming. But what does global warming mean for the Earth? The principal effects of global warming are as follows:

Alterations in Climate Zones

Why is there such concern about this increase in atmospheric temperature? The climate system is a very complex system and is interrelated with other Earth systems. A change in one part of the climate system will impact other parts of the Earth system but not necessarily to the same extent. A rise in global temperature could cause changes in the precipitation pattern, bringing drought to some areas and excessive rain and floods to others. This would also disrupt agriculture in many regions causing famines and death due to starvation. Most of the nation's corn and wheat is grown in the central part of the US. This area is likely to experience a decrease in precipitation causing crop production to decrease. At the same time, some areas in Canada that now experience short growing seasons and little rainfall might experience longer growing seasons and more rainfall thus increasing their crop production.

Human misuse of land is causing formerly productive land to become deserts. This is known as desertification.

Desertification

Over grazing of domestic animals, cutting trees, and bad farming practices are creating over 27 thousand square miles of new desert land each year. Desertification is causing dry but habitable land to become uninhabitable. In many marginally habitable areas, such as the Sahel or southern flank of the Sahara Desert, nomadic tribes farm areas with little or no emphasis on sustainability of the soil, until the soil becomes incapable of growing crops. At that time, many of these tribal peoples simply pack up and move to a nearby marginally habitable area and the practice begins all over again.

Because deserts are by definition arid areas, and the farmed out soil cannot sustain any plant growth, soil erosion becomes an ever-increasing fact of life. As the top soil is blown or washed away and no plants are grown to rejuvenate the soil, the soil profile becomes permanently degraded, with little or no hope of ever becoming productive again.

SEA LEVEL CHANGES

Climate warming would cause some of the polar ice and glaciers to melt causing sea levels to rise. Rising ocean levels would flood low lying areas causing millions of people to flee to higher ground. Salt water incursion, the process by which saline water would invade coastal aquifers and rivers causing the water to be unusable for drinking, irrigation, and industrial users. Polar ice caps and glaciers reflect a large amount of the Sun's energy back to space. If these decrease in size it would allow more of the energy to be absorbed by the surface contributing further to global warming.

Plant and animal species that now inhabit our coastal zones will also be affected by sea level changes. Estuaries, salt and fresh water marshes, tidal lands and other coastal wetlands would migrate landward, and as they do, geographic changes in areas of agricultural productivity and the homes of various plants and animals in the coastal food chains would take place. The availability of food, water, and shelter will diminish and those species that are not able to adapt will become extinct.

What is being done or can be done to address the issue of global warming. Clearly, the one thing that we can do is to reduce the amount of green house gases into the atmosphere.

Reduction of Greenhouse Gas Emissions

As we have seen, the burning of coal, oil, natural gas, and wood increases the amount of greenhouse gases in the atmosphere. In addition, other activities by man have contributed to global warming. There is no doubt that the demands of our growing population has put severe stresses on the Earth systems. What is being done or should be done about global warming? The 1992 Earth Summit in Rio de Janeiro called for nations to report on their releases of greenhouse gases and their efforts to reduce these gases. Over 50 nations have joined this effort since 1994.

Kyoto Protocol

The 1997 Kyoto Conference now calls for the United States and all other developed nations to agree to reduce the amount of greenhouse gas emissions into the atmosphere to a level 15% below the 1990 level, by the year 2010.

The principal reason for burning fossil fuel is to produce energy. An obvious first step in reducing greenhouse gas emissions into the atmosphere is to reduce energy consumption. While all nations and most individuals waste energy, the people of the United States are the most wasteful. Reduction of energy consumption by reduction of energy waste is a practice which is relatively easy to accomplish, and one which will not cause us any undue hardships. Beyond that, it is absolutely necessary that we (in the United States and worldwide) must begin the process of transition from using fossil fuels as our primary source to using non-greenhouse gas producing energy sources. The transition to the use of alternative energy sources is going to be expensive in the short run, but must be encouraged and funded.

Upper Atmosphere Ozone Layer

The stratosphere is that layer of the atmosphere that extends from 17 km to 48 km above sea level. Within the lower parts of the stratosphere there exists a "layer" rich in ozone gas. Ozone, whose chemical formula is O3, is important because it absorbs ultraviolet radiation from the sun, and thus limits the amount of that radiation which reaches the earth's surface. Recent atmospheric studies show that the amount of ozone in the upper atmosphere varies from one location to another, and from one time to the next. This means that the amount of ultraviolet radiation reaching the earth also varies.

This is important because ultraviolet radiation is known to be harmful to many living things, including humans. Any reduction in the amount of ozone in the upper atmosphere can have detrimental effects. It has been well documented in the past few decades that there is a roughly circular area in the ozone layer where the ozone concentration is dangerously low. This hole in the ozone layer is not constant in size or location, but it appears to be most problematic in the Southern Hemisphere, over Antarctica, Australia and New Zealand. Unlike many environmental issues, where we do not know for sure the causes and effects of the problem, in this case we do.

One of the primary causes for the reduction of ozone in the upper atmosphere is an increase in a series of man-made chlorine-and fluorine-compounds known as chlorofluorocarbons.

Chlorofluorocarbons (CFCs)

A description of the properties and uses of chlorofluorocarbons and their effect on the production of ozone is presented in the unit on air pollution. For a discussion of the health effects of the reduction of stratospheric ozone, click here.

For a discussion of what can be done about CFCs, go to the section on Changing Propellants and Refrigerants.

Health Effects

There is a clear and present relationship between UV rays and human health. In short, ultraviolet radiation causes skin cancer and cataracts. There has been a significant increase in number of cases of both skin cancer and cataracts in many parts of both the northern and southern hemispheres in the last 20 years. The relationship between ozone, UV radiation and skin cancer and cataract cases is particularly acute in Australia, New Zealand, South Africa, Argentina and Chile. Australian television now regularly report UV levels and warnings.

Again, in this case, we not only know the cause and effects of the environmental issue under consideration, we also know what we must do to reduce the problem.

Changing Propellants and Refrigerants

If ultraviolet radiation causes skin cancer and cataracts, and if the reduction of ozone causes an increase in the amount of UV radiation reaching the earth, and if the emission of chlorofluorocarbon compounds is a primary contributor to the reduction of ozone, it is clear that we must drastically reduce the emission of CFSs to the atmosphere. One way to effect that change is to reduce the use, where possible, of CFCs.

Fortunately, there are other compounds that can be effectively used for propellants. Unfortunately, there are few compounds that can be effectively substituted for freon used n refrigerants. Recycling of refrigerant can reduce the problem. But, the simplest solutions are to reduce your exposure to UV radiation, particularly in those areas where the UV levels are highest, and to stop using products that include CFCs. That will not be easy, because of our affinity for sun tanned skin and because many CFC-bearing products are inexpensively available and effective. However, while there will be some short-term cost increase of substitute products and some short-term loss of comfort, not changing our destructive actions will cause an existing problem to become significantly more severe.

The Montreal Protocol on Substances that Deplete the Ozone Layer (1987 and 1990) has been signed by over 70 nations. This Protocol calls for the reduction in the emission of CFCs (but not other ozone-depleting substances) into the atmosphere by 5% by the year 2000.

In 1992, representatives of most nations met in Copenhagen and agreed to: (1) phase out production of CFCs and other ozone-depleting substances by 1996; and (2) freeze consumption of CFCs at 1991 levels by 1996.

These treaties and protocols are working. By 1992, global CFC production was one-half of the production in 1988, the historic high, and down to approximately the same level as in 1972.

Ozone Depletion

Ozone is a pungent, slightly bluish gas. The ozone molecule is made up of three oxygen atoms. The oxygen molecule required for human respiration is comprised of two oxygen atoms. Ozone is formed when the two atoms of an oxygen molecule are separated as a result of ultraviolet (UV) radiation from the sun. The individual atoms of oxygen combine with the oxygen molecules to form ozone.

The Ozone Layer

The Ozone Layer is a concentration of ozone molecules in the stratosphere. The layer of the earth's atmosphere that surrounds us is called the troposphere. The stratosphere, the next higher layer, extends about 10-50 kilometers above the earth's surface.

Good and Bad Ozone: About 90% of the Earth's ozone is in the stratosphere. Ozone in the stratosphere could be called good ozone because it shields the earth from ultraviolet radiation. The remaining 10% of ozone is in the troposphere, where at certain concentrations it could be harmful to public health.

About Good Ozone: Ozone absorbs Ultra-Violet radiation. UV radiation from the sun can damage DNA in living systems. This radiation, called UV-B, occurs at wavelengths between 280 and 320 nanometers (one nanometer is equal to one billionth of a meter). Ozone is continuously being destroyed as it absorbs UV-B. However, the absorption process produces atomic oxygen that reacts with molecular oxygen to rebuild ozone.

CFC: Chlorofluorocarbon

The "wonder chemical" CFC was discovered in the 1930's and commercialized in the 1940's. CFC which was used as a refrigerant or as a cleaning solvent, was relatively inexpensive, highly effective, stable in the atmosphere and nontoxic to humans. These characteristics lead to its widespread use throughout broad sectors of modern society.

CFC is nontoxic, noncorrosive, nonflamable, and extremely stable. Some CFC sources are:

- Rigid Foams, used for insulation and packaging
- Flexible Foams, used in furniture, bedding, car seats
- Air Conditioners-automobile, room
- Aerosols-propellants in aerosol sprays

Too good to be true!: When released in the atmosphere, CFC's travel up to the stratosphere, where it reacts with UV radiation resulting in the release of chlorine. The chlorine acts as a catalyst in the reaction that results in the repeated destruction of ozone molecules.

Main Hurdle

As you will recall the ozone molecules are made up of three oxygen atoms. The oxygen molecule is made up of two oxygen atoms. When the CFCs react with the UV radiation it releases chlorine. The chlorine reacts with the ozone resulting in the production of one oxygen molecule and one chlorine monoxide molecule. The chlorine monoxide molecule then can combine with an oxygen atom, forming an oxygen molecule and freeing the chlorine to repeat the process all over again.

The Ozone Hole

In 1985 the British Antartic Survey reported that the ozone layer over Antartica had shrunk each year since the late 1970's. In 1998 the ozone hole over Antartica covered an area of about 10 million square miles. By comparison the surface area of North America is about 8.1 million square miles. After the winter of 1999-2000, NASA and a team of European scientists observed that in the Artic about 60% of stratospheric ozone was lost.

The Dobson Unit: DU measures the total amount of ozone in a column of air from ground level to the top of the atmosphere, based on spectral analysis of absobed UV light. A typical reading for a healthy polar ozone layer might be in the 300-450 DU range. Consistent measurements have shown a steady decline in the Ozone Shield. The readings have dipped below 100 DU in Antartica.

Environmental Protection Agency-EPA

In addition to regulating the end of production of the ozone-depleting substances (ODS), the U.S. Environmental Protection Agency (EPA) implements several other programs to protect the ozone layer. These programs include refrigerant recycling, product labeling, banning nonessential uses of certain compounds, and reviewing substitutes.

HEALTH AND ENVIRONMENTAL IMPACT

- Premature skin aging
- Skin Cancer
- Suppression of the Human Immune Response System
- Damage to the Human Eye
- Damage to Crops and Other Terrestrial Ecosystems
- Damage to Certain Marine Organisms

Before flue gas desulfurization was installed, the emissions from this power plant in New Mexico contained excessive amounts of sulfur dioxide.

Air pollution is a chemical, physical (e.g. particulate matter), or biological agent that modifies the natural characteristics of the atmosphere. The atmosphere is a complex, dynamic natural gaseous system that is essential to support life on planet earth. Stratospheric ozone depletion due to air pollution has long been recognized as a threat to human health as well as to the earth's ecosystems.

Worldwide air pollution is responsible for large numbers of deaths and cases of respiratory disease. Enforced air quality standards, like the Clean Air Act in the United States, have reduced the presence of some pollutants. While major stationary sources are often identified with air pollution, the greatest source of emissions is actually mobile sources, principally the automobile. Gases such as carbon dioxide, which contribute to global warming, have recently gained recognition as pollutants by some scientists. Others recognise the gas as being essential to life, and therefore incapable of being classed as a pollutant.

Deaths

The World Health Organization estimates that 4.6 million people die each year from causes directly attributable to air pollution. Many of these mortalities are attributable to indoor air pollution. Worldwide more deaths per year are linked to air pollution than to automobile accidents. Research published in 2005 suggests that 310,000 Europeans die from air pollution annually. Direct causes of air pollution related deaths include aggravated asthma, bronchitis, emphysema, lung and heart diseases, and respiratory allergies. The US EPA estimates that a proposed set of changes in diesel engine technology (Tier 2) could result in 12,000 fewer premature mortalities, 15,000 fewer heart attacks, 6,000 fewer emergency room visits by children with asthma, and 8,900 fewer respiratory-related hospital admissions each year in the United States.

The worst short term civilian pollution crisis in India was the 1984 Bhopal Disaster. Leaked industrial vapors from the Union Carbide factory, belonging to Union Carbide, Inc., U.S.A., killed more than 2,000 people outright and injured anywhere from 150,000 to 600,000 others, some 6,000 of whom would later die from their injuries. The United Kingdom suffered its worst air pollution event when the December 4th Great Smog of 1952 formed over London. In six days more than 4,000 died, and 8,000 more died within the following months. An accidental leak of anthrax spores from a biological warfare laboratory in the former USSR in 1979 near Sverdlovsk is believed to have been the cause of hundreds of civilian deaths. The worst single incident of air pollution to occur in the United States of America occurred in Donora, Pennsylvania in late October, 1948, when 20 people died and over 7,000 were injured.

Sources

Air pollutants are classified as either directly released or formed by subsequent chemical reactions. A direct release air pollutant is one that is emitted directly from a given source, such as the carbon monoxide or sulfur dioxide, all of which are byproducts of combustion; whereas, a subsequent air pollutant is formed in the atmosphere through chemical reactions involving direct release pollutants. The formation of ozone in photochemical smog is the most important example of a subsequent air pollutant. Using a controlled burn on a field in South Georgia in preparation for spring planting.

- Combustion-fired power plants (coal-fired power plants contribute 40% of world's sulphur and mercury)
- Controlled burn practices used in agriculture and forestry management
- Motor vehicles generating air pollution emissions.
- Marine vessels, such as container ships or cruise ships, and related port air emissions
- Burning fossil fuels
- Burning wood, fireplaces, stoves, furnaces and incinerators

Other anthropogenic sources:

- Oil refining, power plant operation and industrial activity in general.

- Chemicals, dust and crop waste burning in farming.
- Fumes from paint, varnish, aerosol sprays and other solvents.
- Waste deposition in landfills, which generate methane.
- Military uses, such as nuclear weapons, toxic gases, germ warfare and rocketry.

Natural Sources

- Dust from natural sources, usually large areas of land with little or no vegetation.
- Methane, emitted by the digestion of food by animals, for example cattle.
- Pine trees, which emit volatile organic compounds (VOCs).
- Radon gas from radioactive decay within the Earth's crust.
- Smoke and carbon monoxide from wildfires.
- Volcanic activity, which produce sulfur, chlorine, and ash particulates.

AIR QUALITY STANDARDS

Canada

In Canada, air quality standards are typically determined by a federal-provincial process known as the Canadian Council of Ministers for the Environment (CCME). CCME created the Canada-Wide Standards for ozone and fine particulate matter (PM 2.5). National Ambient Air Quality Objectives also exist for criteria air contaminants, although these are used less frequently than the CCME standards.

United Kingdom

Managing air quality in the United Kingdom is the responsibility of DEFRA. DEFRA core purpose is to improve the current and future quality of life a broad range of responsibilities of which environmental protection and air quality are key.

The present secretary for state the Rt Honorable David Milliband said in a recent speech: If citizens, businesses and nations are to change their behaviour, they must be confident that their actions-will be reciprocated. Citizens need to know their neighbours are committed and that together their actions will have critical mass. Business need to know that the bar is being raised, but it is being raised for all businesses. Nations need to know that others will follow suit.

This succinctly summarises the reason why the UK is harmonising its air quality strategy with the European Union air pollution strategy; so that communities that commit to improving air quality will not be penalised economically by becoming less competitive.

Air quality targets set by DEFRA are mostly aimed at local government representatives responsible for the management of air quality in cities, where air quality management is the most urgent. The UK has established an air quality network where levels of the key air pollutants are published by monitoring centers. Air quality in Oxford, Bath and London is particuarly poor. One controversial study created by Calor Gas and published in the Guardian newspaper compared walking in Oxford on an average day to smoking over sixty light cigarettes.

'Cigarette equivalents' is obviously a headline capturing measure and more reliable and accepted comparison data can be collected from UK Air Quality Archive which allows the user to compare a cities management of pollutants against objectives set by DEFRA in 2000.

It is, however, important to evaluate several aspects of air pollution, and especially to take into consideration overall average values, rather than localized peak values sometimes cited. The UK National Air Quality Information Archive offers almost real-time monitoring "current maximum" air pollution measurements for many UK towns and cities. This source offers a wide range of constantly updated data, including:

Hourly Mean Ozone (μg/m^3);

Hourly Mean Nitrogen dioxide (μg/m^3);

max 15 min mean Sulphur dioxide (μg/m^3);

8 Hourly Mean Carbon monoxide (mg/m^3), and

24 Hour mean PM10 Particles (μg/m^3 Grav Equiv)

DEFRA acknowledges that air pollution has a significant effect on health and has produced a simple banding system that is used to create a daily warning system that is issued by the BBC Weather service to indicate air pollution levels. DEFRA has published guidelines for people suffering from respiratory and heart diseases.

United States

Looking down from the Hollywood Hills, with Griffith Observatory on the hill in the foreground, air pollution is visible in downtown Los Angeles on a late afternoon.

In the 1960s, 70s, and 90s, the United States Congress enacted a series of Clean Air Acts which significantly strengthened regulation of air pollution. Individual U.S. states, some European nations and eventually the European Union followed these initiatives. The Clean Air Act sets numerical limits on the concentrations of a basic group of air pollutants and provides reporting and enforcement mechanisms.

In 1999, the United States EPA replaced the Pollution Standards Index (PSI) with the Air Quality Index (AQI) to incorporate new PM2.5 and Ozone standards.

Atmospheric dispersion modeling

The basic technology for analyzing air pollution is through the use of a variety of mathematical models for predicting the transport of air pollutants in the lower atmosphere. The principal methodologies are:

- Point source dispersion, used for simple industrial sources.
- Line source dispersion, used for airport and roadway air dispersion modeling
- Area source dispersion, used for forest fires or duststorms
- Photochemical models, used to analyze reactive pollutants such as form smog

The point source problem is the best understood, since it involves simpler mathematics and has been studied for a long period of time, dating back to about the year 1900. It uses a Gaussian dispersion model to forecast the air pollution isopleths, with consideration given to wind velocity, stack height, emission rate, stability class (a measure of atmospheric turbulence). This model has been extensively validated and calibrated with experimental data for all sorts of atmospheric conditions.

The roadway air dispersion model was developed starting in the late 1950s and early 1960s in response to requirements of the National Environmental Policy Act and the U.S. Department of Transportation (then known as the Federal Highway Administration) to understand impacts of proposed new highways upon air quality, especially in urban areas. Several research groups were active in this model development, among which were: the Environmental Research and Technology (ERT) group in Lexington, Massachusetts, the ESL Inc. group in Sunnyvale, California and the California Air Resources Board group in Sacramento, California. The research of the ESL group received a boost with a contract award from the U.S. Environmental Protection Agency to validate a line source model using sulfur hexafluoride as a tracer gas. This program was successful in validating the line source model developed by ESL inc. Some of the earliest uses of the model were in court cases involving highway air pollution, the Arlington, Virginia portion of Interstate 66 and the New Jersey Turnpike widening project through East Brunswick, New Jersey.

Area source models were developed in 1971 through 1974 by the mid 1970s by the ERT and ESL groups, but addressed a smaller fraction of total air pollution emissions, so that their use and need was not as widespread as the line source model, which enjoyed hundreds of different applications as early as the 1970s. Similarly photochemical models were developed primarily in the 1960s and 1970s, but their use was more specialized and for regional needs, such as understanding smog formation in the Livermore Valley, California.

Indoor Air Pollution, or Indoor Air Quality (IAQ)

The lack of ventilation indoors concentrates air pollution where people have greatest exposure times. Radon (Rn) gas, a carcinogen, is exuded from the Earth in certain locations and trapped inside houses. Researchers have found that radon gas is responsible for over 1,800 deaths annually in the United Kingdom.

Building materials including carpeting and plywood emit formaldehyde (H_2CO) gas. Paint and solvents give off volatile organic compounds (VOCs) as they dry. Lead paint can degenerate into dust and be inhaled. Intentional air pollution is introduced with the use of air fresheners, incense, and other scented items. Controlled wood fires in stoves and fireplaces can add significant amounts of smoke particulates into the air, inside and out. Indoor air pollution may arise from such mundane sources as shower water mist containing arsenic or manganese, both of which are damaging to inhale. The arsenic (As^{3+}) can be trapped with a shower nozzle filter.

Indoor pollution fatalities may be caused by using pesticides and other chemical sprays indoors without proper ventilation, and many homes have been destroyed by accidental pesticide explosions.

Carbon monoxide (CO) poisoning is a quick and silent killer, often caused by faulty vents and chimneys, or by the burning of charcoal indoors. 56,000 Americans died from CO in the period 1979-1988. Chronic carbon monoxide poisoning can result even from poorly adjusted pilot lights. Smoke inhalation is a common cause of death in victims of house fires. Traps are built into all domestic plumbing to keep deadly sewer gas, hydrogen sulfide, out of interiors. Clothing emits tetrachloroethylene, or other dry cleaning fluids, for days after dry cleaning.

Though its use has now been banned in many countries, the extensive use of asbestos in industrial and domestic environments in the past has left a potentially very dangerous material in many localities. Asbestosis is a chronic inflammatory medical condition affecting the tissue of the lungs.

It occurs after long-term, heavy exposure to asbestos, e.g. in mining or in the installation or removal of asbestos-containing materials from structures. Sufferers have severe dyspnea (shortness of breath) and are at an increased risk regarding several different types of lung cancer. As clear explanations are not always stressed in non-technical literature, care should be taken to distinguish between several forms of relevant diseases. According to the World Health Organisation (WHO), these may defined as; asbestosis, lung cancer, and mesothelioma (generally a very rare form of cancer, when more widespread it is almost always associated with prolonged exposure to asbestos).

Biological sources of air pollution are also found indoors, as gases and airborne particulates. Pets produce dander, people produce dust from minute skin flakes, dust mites in bedding, carpeting and furniture produce enzymes and micron-sized fecal droppings, inhabitants emit methane, mold forms in walls and generates mycotoxins and spores, air conditioning systems can incubate Legionnaires' disease and mold, toilets can emit feces-tainted mists, and houseplants, soil and surrounding gardens can produce pollen, dust, and mold. Indoors, the lack of air circulation allows these airborne pollutants to accumulate more than they would otherwise occur in nature.

Reduction efforts

There are many available air pollution control technologies and urban planning strategies available to reduce air pollution; however, worldwide costs of addressing the issue are high. Enforced air quality standards, like the Clean Air Act in the United States, have reduced the presence of some pollutants.

Many countries have programs to or are debating how to reduce dependence on fossil fuels for energy production and shift toward renewable energy technologies or nuclear power plants.

Efforts to reduce pollution from mobile sources includes primary regulation (many developing countries have permissive regulations), expanding regulation to new sources (such as cruise and transport ships, farm equipment, and small gas-powered equipment such as lawn trimmers, chainsaws, and snowmobiles), increased fuel efficiency (such as through the use of hybrid vehicles), conversion to cleaner fuels (such as bioethanol, biodiesel), or conversion to electric vehicles with renewable energy sources (batteries or clean fuel such as hydrogen being used for transport and storage).

Control Devices

The following items are commonly used as pollution control devices by industry or transportation devices. They can either destroy contaminants or remove them from an exhaust stream before it is emitted into the atmosphere.

- Particulate control
  - o Mechanical collectors (dust cyclones, multicyclones)
  - o Electrostatic precipitators
  - o Fabric filters (baghouses)
  - o Particulate scrubbers
- NOx control
  - o Low NOx burners
  - o Selective catalytic reduction (SCR)
  - o Selective non-catalytic reduction (SNCR)
  - o NOx scrubbers
  - o Exhaust gas recirculation
  - o Catalytic converter (also for VOC control)
- VOC abatement
  - o Adsorption systems, such as activated carbon
  - o Flares
  - o Thermal oxidizers
  - o Catalytic oxidizers
  - o Biofilters
  - o Absorption (scrubbing)
  - o Cryogenic condensers
- Acid Gas/SO_2 control
  - o Wet scrubbers
  - o Dry scrubbers
  - o Flue gas desulfurization
- Mercury control
  - o Sorbent Injection Technology

  - o Electro-Catalytic Oxidation (ECO)
  - o K-Fuel
- Dioxin and furan control
- Ambient cleaning systems
- Associated equipment
  - o Source capturing systems
  - o Continuous emissions monitoring systems (CEMS)

Greenhouse Effect and Ocean Acidification

The greenhouse effect is a phenomenon whereby carbon dioxide levels are thought to create a condition in the upper atmosphere, causing a trapping of excess heat and leading to increased surface temperatures. It shares this property with many other gases, and water vapour produces a larger effect than carbon dioxide. Other greenhouse gases include methane and NOx.

This effect has been understood by scientists for about a century, and technological advancements during this period have helped increase the breadth and depth of data relating to the phenomenon. A number of studies have investigated the potential for long-term rising levels of atmospheric carbon dioxide to cause slight increases in the acidity of ocean waters and the possible effects of this on marine ecosystems. However, carbonic acid is a very weak acid, and is utilised by marine organisms during photosynthesis.

Acid Rain

'Acid rain' (or more accurately acid precipitation) occurs when sulfur dioxide and nitrogen oxides are emitted into the atmosphere, undergo chemical transformations and are absorbed by water droplets in clouds. The droplets then fall to earth as rain, snow, mist, dry dust, hail, or sleet. This can increase the acidity of the soil, and affect the chemical balance of lakes and streams.

The term "acid rain" is sometimes used more generally to include all forms of acid deposition-both wet deposition, where acidic gases and particles are removed by rain or other precipitation, and dry deposition removal of gases and particles to the Earth's surface in the absence of precipitation. Acid rain is defined as any type of precipitation with a pH that is unusually low. Dissolved carbon dioxide-dissociates to form weak carbonic acid giving a pH of approximately 5.6 at typical atmospheric concentrations of CO_2.

Therefore a pH of less than 5.6 has sometimes been used as a definition of acid rain. However, natural sources of acidity mean that in remote areas, rain has a pH which is between 4.5 and 5.6 with an average value of 5.0 and so rain with a pH of less than 5 is a more appropriate definition. The US EPA says, "Acid rain is a serious environmental problem that affects large parts of the US and Canada" Acid rain accelerates weathering in carbonate rocks and accelerates building weathering. It also contributes to acidification of rivers, streams, and forest damage at high elevations. When the acid builds up in rivers and streams it can kill fish.

History and Trends

Evidence for an increase in the levels of acid rain comes from analyzing layers of glacial ice. These show a sudden decrease in pH from the start of the industrial revolution of 6 to 4.5 or 4. Other information has been gathered from studying organisms known as diatoms which inhabit ponds. Over the years these die and are deposited in layers of sediment on the bottoms of the ponds. Diatoms thrive in certain pHs, so the numbers of diatoms found in layers of increasing depth give an indication of the change in pH over the years.

Since the industrial revolution, emissions of sulfur and nitrogen oxides to the atmosphere have increased. Occasional pH readings of well below 2.4 (the acidity of vinegar) have been reported in industrialized areas. Industrial acid rain is a substantial problem in the People's Republic of China, Eastern Europe, Russia and areas down-wind from them. These areas all burn sulfur-containing coal to generate heat and electricity. The problem of acid rain not only has increased with population and industrial growth, but has become more widespread.

The use of tall smokestacks to reduce local pollution has contributed to the spread of acid rain by releasing gases into regional atmospheric circulation. Often deposition occurs a considerable distance downwind of the emissions, with mountainous regions tending to receive the most (simply because of their higher rainfall). An example of this effect is the low pH of rain (compared to the local emissions) which falls in Scandinavia.

Acid rain was first reported in Manchester, England, which was an important city during the Industrial Revolution. In 1852, Robert Angus Smith found the relationship between acid rain and atmospheric pollution. The term "acid rain" was used for the first time by him in 1872. Though acid rain was discovered in 1852, it wasn't until the late 1960s that scientists began widely observing and studying the phenomenon. Canadian Harold Harvey was among the first to research a "dead" lake. Public awareness of acid rain in the U.S increased in the 1990s after the New York Times promulgated reports from the Hubbard Brook Experimental Forest in New Hampshire of the myriad deleterious environmental effects demonstrated to result from it.

EMISSIONS OF CHEMICALS LEADING TO ACIDIFICATION

Volcanic Injection

The most important gas which leads to acidification is sulfur dioxide. Emissions of nitrogen oxides which are oxidised to form Nitric acid are of increasing importance due to stricter controls on emissions of sulfur containing compounds. 70 Tg(S) per year in the form of SO_2 comes from fossil fuel combustion and industry, 2.8 Tg(S) from wildfires and 7-8 Tg(S) per year from volcanoes.

Natural Emissions

The principal natural phenomena that contribute acid-producing gases to the atmosphere are emissions from volcanoes and those from biological processes that occur on the land, in wetlands, and in the oceans. The major biological source of sulfur containing compounds is dimethyl sulfide. The effects of acidic deposits have been detected in glacial ice thousands of years old in remote parts of the globe.

Deposition

Wet Deposition

Wet deposition of acids occurs when any form of precipitation (rain, snow, etc.) removes acids from the atmosphere and delivers it to the Earth's surface. This can result from the deposition of acids produced in the raindrops or by the precipitation removing the acids either in clouds or below clouds. Wet removal of both gases and aerosol are both of importance for wet deposition.

Dry Deposition

Acid deposition also occurs via dry deposition in the absence of precipitation. This can be responsible for as much as 20 to 60% of total acid deposition. This occurs when particles and gases stick to the ground, plants or other surfaces.

Adverse Effects

Acid rain has been shown to have adverse impacts on forests, freshwaters and soils, killing off insect and aquatic lifeforms as well as causing damage to buildings and having possible impacts on human health.

Surface Waters and Aquatic Animals

Both the lower pH and higher aluminium concentrations in surface water that occur as a result of acid rain can cause damage to fish and other aquatic animals. At pHs lower than 5 most fish eggs will not hatch and lower pHs can kill adult fish. As lakes become more acidic biodiversity is reduced. Acid rain has eliminated insect life and some fish species, including the brook trout in some Appalachian streams and creeks.

Soils

Soil biology can be seriously damaged by acid rain. Some tropical microbes can quickly consume acids but other microbes are unable to tolerate low pHs and are killed. The enzymes of these microbes are denatured (changed in shape so they no longer function) by the acid. The hydronium ions of acid rain also mobilize toxins and leach away essential nutrients and minerals.

Forests and other Vegetation

Acid rain can slow the growth of forests, cause leaves and needles to turn brown and fall off and die. In extreme cases trees or whole areas of forest can die. The death of trees is not usually a direct result of acid rain, often it weakens trees and makes them more susceptible to other threats. Damage to soils can also cause problems. High altitude forests are especially vulnerable as they are often surrounded by clouds and fog which are more acidic than rain. Other plants can also be damaged by acid rain but the effect on food crops is minimised by the application of fertilizers to replace lost nutrients. In cultivated areas, limestone may also be added to increase the ability of the soil to keep the pH stable, but this tactic is largely unusable in the case of wilderness lands. Acid Rain depletes minerals from the soil and then it stunts the growth of the plant.

Human Health

Some scientists have suggested direct links to human health, but none have been proven.. However, fine particles, a large fraction of which are formed from the same gases as acid rain (sulfur dioxide and nitrogen dioxide), have been shown to cause illness and premature deaths.

Other Adverse Effects

Acid rain can also cause damage to certain building materials and historical monuments. Acid rain can cause erosion on ancient and valuable statues and has caused considerable damage. This is because the sulfuric acid in the rain chemically reacts with the calcium in the stones (limestone, sandstone, marble and granite) to create gypsum, which then flakes off. This is also commonly seen on old gravestones where the acid rain can cause the inscription to become completely illegible. Acid rain also causes an increased rate of oxidation for iron. Visibility is also reduced by sulfate and nitrate in the atmosphere.

Prevention Methods

In the United States, many coal-burning power plants use Flue gas desulfurization (FGD) to remove sulfur-containing gases from their stack gases. An example of FGD is the we scrubber which is commonly used in the U.S. and many othe countries. A wet scrubber is basically a reaction tower equippe with a fan that extracts hot smoky stack gases from a powe plant into the tower. Lime or limestone in slurry form is als injected into the tower to mix with the stack gases and combin with the sulfur dioxide present. The calcium carbonate of th limestone produces pH-neutral calcium sulfate that is physicall removed from the scrubber. That is, the scrubber turns sulfu pollution into industrial sulfates. In some areas the sulfates ar sold to chemical companies as gypsum when the purity (calcium sulfate is high. In others, they are placed in landfil However, the effects of acid rain can last for generations, a the effects of ph level change can stimulate the continue leaching of undesirable chemicals into otherwise pristine wat sources, killing off vulnerable insect and fish species and blockir efforts to restore native life.

International Treaties

A number of international treaties on the long ran transport of atmospheric pollutants have been agreed e. Sulphur Emissions Reduction Protocol and Convention on Lon Range Transboundary Air Pollution.

Emissions Trading

An even more benign regulatory scheme involves emissio trading. In this scheme, every current polluting facility is giv an emissions license that becomes part of capital equipme Operators can then install pollution control equipment, a sell parts of their emissions licenses. The main effect of th is to give operators real economic incentives to install polluti controls. Since public interest groups can retire the licenses purchasing them, the net result is a continuously decreasi and more diffused set of pollution sources. At the same tin no particular operator is ever forced to spend money witho a return of value from commercial sale of assets.

AIR QUALITY INDEX

The Air Quality Index (AQI) is a standardized indicato the air quality in a given location. It measures mainly grou level ozone and particulates (except the pollen count), but may also include sulphur dioxide, and nitrogen dioxide. Various agencies around the world measure such indices, though definitions may change between places.

The United States Environmental Protection Agency (EPA) and the Meteorological Service of Canada (MSC) differ on what AQI structure and health classification is used:

Health classifications used by the EPA:

- 0-50 Good is usually green
- 51-100 Moderate is usually yellow
- 101-150 Unhealthy for sensitive groups is usually orange
- 151-200 Unhealthy is usually red
- 201-300 Very unhealthy is usually purple
- 301-500 Hazardous is usually maroon

(AQI 100 corresponds to 0.08 ppm ozone; other levels for other pollutants) EPA

Health classifications used by the MSC:

- 0-25*: Good (green)
- 26*-50: Moderate (yellow)
- 51-100: Poor (orange/red)
- 101+: Very poor (purple)

In Ontario, 31 is the upper limit for good and 32 the lower limit for moderate. The AQI can worsen (go up) due to lack of dilution with fresh air. Stagnant air, often caused by an anticyclone or temperature inversion, or other lack of winds lets air pollution remain in a local area. On these days, the news media may ask the public to carpool or use public transport, or take other air pollution prevention measures such as teleworking.

Other Indexes

United States

The United States Environment Protection Agency (USEPA) developed the Pollutant Standards Index (PSI) to provide accurate, timely and easily understandable information about daily levels of air pollution. It is no longer in use, having been replaced by the AQI, which is more sensitive. For example, particulate matter with an aerodynamic diameter less than 2.5 micrometres (PM2.5) is a sub index, replacing the less sensitive PM10 component of the PSI.

United Kingdom

The Met Office of the United Kingdom (UK) issues air quality forecasts wherein the level of pollution is described either as an index (ranging from 1 to 10) or as a banding (low, moderate, high or very high). These levels are based on the health effects of each pollutant.

Index	Banding	Health Effect
1-3	Low	Effects are unlikely to be noticed even by individuals who know they are sensitive to air pollutants.
4-6	Moderate	Mild effects, unlikely to require action, may be noticed amongst sensitive individuals.
7-9	High	Significant effects may be noticed by sensitive individuals and action to avoid or reduce these effects may be needed (e.g. reducing exposure by spending less time in polluted areas outdoors). Asthmatics will find that their 'reliever' inhaler is likely to reverse the effects on the lung.
10	Very High	The effects on sensitive individuals described for 'High' levels of pollution may worsen.

The forecast is produced for a number of different pollutants and their typical health effects are shown in the following table.

Pollutant	Health Effects at High Level
Nitrogen dioxide Ozone Sulphur dioxide	These gases irritate the airways of the lungs, increasing the symptoms of those suffering from lung diseases.
Particulates	Fine particles can be carried deep into the lungs where they can cause in flammation and a worsening of heart and lung diseases

Hong Kong

The Air Pollution Index (API) levels for Hong Kong are related to the measured concentrations of ambient respirable suspended particulate (RSP), sulphur dioxide (SO_2), carbon monoxide (CO), ozone (O_3) and nitrogen dioxide (NO_2) over a 24-hour period based on the potential health effects of air pollutants. An API level at or below 100 means that the pollutant levels are in the satisfactory range over 24 hour period and pose no acute or immediate health effects. However, air pollution consistently at "High" levels (API of 51 to 100) in a year may mean that the annual Hong Kong "Air Quality Objectives" for protecting long-term health effects could be violated. Therefore, chronic health effects may be observed if one is persistently exposed to an API of 51 to 100 for a long time.

"Very High" levels (API in excess of 100) means that levels of one or more pollutant(s) is/are in the unhealthy range. The Hong Kong Environmental Protection Department provides advice to the public regarding precautionary actions to take for such levels.

API	Air Pollution Level	Health Implications
0-25	Low	Not expected.
26-50	Medium	Not expected for the general population.
51-100	High	Acute health effects are not expected but chronic effects may be observed if one is persistently exposed to such levels.
100-200	Very High	People with existing heart or respiratory illnesses may notice mild aggravation of their health conditions. Generally healthy individuals may also notice some discomfort.
201-500	Very High	People with existing heart or respiratory illnesses may experience significant aggravation of their symptoms. There may also be widespread symptoms in the healthy population (e.g. eye irritation, wheezing, coughing, phlegm and sore throats).

Malaysia

The air quality in Malaysia is reported as the API or Air Pollution Index. Four of the index's pollutant components (i.e., carbon monoxide, ozone, nitrogen dioxide and sulfur dioxide) are reported in ppmv but PM10 particulate matter is reported in $\mu g/m^3$. Unlike the American AQI, the index number can exceed 500. Above 500, a state of emergency is declared in the reporting area. Usually, this means that non-essential government services are suspended, and all ports in the affected area closed. There may also be a prohibition on private sector commercial and industrial activities in the reporting area excluding the food sector.

Atmospheric Dispersion Modeling

Atmospheric dispersion modeling is the mathematical simulation of how air pollutants disperse in the ambient atmosphere. It is performed with computer programs that solve the mathematical equations and algorithms which simulate the pollutant dispersion. The dispersion models are used to estimate or to predict the downwind concentration of air pollutants emitted from sources such as industrial plants and vehicular traffic. Such models are important to governmental agencies tasked with protecting and managing the ambient air quality. The models are typically employed to determine whether existing or proposed new industrial facilities are or will be in compliance with the National Ambient Air Quality Standards (NAAQS) in the United States and other nations. The models also serve to assist in the design of effective control strategies to reduce emissions of harmful air pollutants.

The dispersion models require the input of data which includes:

- Meteorological conditions such as wind speed and direction, the amount of atmospheric turbulence (as characterized by what is called the "stability class"), the ambient air temperature and the height to the bottom of any inversion aloft that may be present.
- Emissions parameters such as source location and height, source vent stack diameter and exit velocity, exit temperature and mass flow rate.

- Terrain elevations at the source location and at the receptor location.
- The location, height and width of any obstructions (such as buildings or other structures) in the path of the emitted gaseous plume.

Many of the modern, advanced dispersion modeling programs include a pre-processor module for the input of meteorological and other data, and many also include a post-processor module for graphing the output data and/or plotting the area impacted by the air pollutants on maps.

The atmospheric dispersion models are also known as atmospheric diffusion models, air dispersion models, air quality models, and air pollution dispersion models.

Gaussian Air Pollutant Dispersion Equation

The technical literature on air pollution dispersion is quite extensive and dates back to the 1930's and earlier. One of the early air pollutant plume dispersion equations was derived by Bosanquet and Pearson. Their equation did not assume Gaussian distribution nor did it include the effect of ground reflection of the pollutant plume.

Sir Graham Sutton derived an air pollutant plume dispersion equation in 1947 which did include the assumption of Gaussian distribution for the vertical and crosswind dispersion of the plume and also included the effect of ground reflection of the plume.

Under the stimulus provided by the advent of stringent environmental control regulations, there was an immense growth in the use of air pollutant plume dispersion calculations between the late 1960s and today. A great many computer programs for calculating the dispersion of air pollutant emissions were developed during that period of time and they were called "air dispersion models".

It should be noted that sz and sy are functions of the atmospheric stability class (i.e., a measure of the turbulence in the ambient atmosphere) and of the downwind distance to the receptor. The two most important variables affecting the degree of pollutant emission dispersion obtained are the height of the emission source point and the degree of atmospheric turbulence. The more turbulence, the better the degree of dispersion.

The resulting calculations for air pollutant concentrations are often expressed as an air pollutant concentration contour map in order to show the spatial variation in contaminant levels over a wide area under study. In this way the contour lines can overlay sensitive receptor locations and reveal the spatial relationship of air pollutants to areas of interest.

The Briggs Plume Rise Equations

The Gaussian air pollutant dispersion equation (discussed above) requires the input of H which is the pollutant plume's centerline height above ground level-and H is the sum of Hs (the actual physical height of the pollutant plume's emission source point) plus sH (the plume rise due the plume's buoyancy).

To determine sH, many if not most of the air dispersion models developed and used between the late 1960s and the early 2000s used what are known as "the Briggs equations." G.A. Briggs published his first plume rise model observations and comparisons in 1965.

In 1968, at a symposium sponsored by CONCAWE (a Dutch organization), he compared many of the plume rise models then available in the literature. In that same year, Briggs also wrote the section of the publication edited by Slade dealing with the comparative analyses of plume rise models. That was followed in 1969 by his classical critical review of the entire plume rise literature, in which he proposed a set of plume rise equations which have became widely known as "the Briggs equations". Subsequently, Briggs modified his 1969 plume rise equations in 1971 and in 1972.

Briggs divided air pollution plumes into these four general categories:

- Cold jet plumes in calm ambient air conditions
- Cold jet plumes in windy ambient air conditions
- Hot, buoyant plumes in calm ambient air conditions
- Hot, buoyant plumes in windy ambient air conditions.

Briggs considered the trajectory of cold jet plumes to be dominated by their initial velocity momentum, and the trajectory of hot, buoyant plumes to be dominated by their buoyant momentum to the extent that their initial velocity momentum was relatively unimportant. Although Briggs proposed plume rise equations for each of the above plume categories, it is important to emphasize that "the Briggs equations" which become widely used are those that he proposed for bent-over, hot buoyant plumes.

In general, Briggs's equations for bent-over, hot buoyant plumes are based on observations and data involving plumes from typical combustion sources such as the flue gas stacks from steam-generating boilers burning fossil fuels in large power plants. Therefore the stack exit velocities were probably in the range of 20 to 100 ft/s (6 to 30 m/s) with exit temperatures ranging from 250 to 500 °F (120 to 260 °C).

Other Types of Pollution

Since the late 1970s the use of CFCs has been heavily regulated because of its destructive effects on the ozone layer. After the development of his atmospheric CFC detector, James Lovelock was the first to detect the presence of CFC's in the air, finding a concentration of 60 parts per trillion of CFC-11 over Ireland. In a self-funded research expedition ending in 1973, Lovelock went on to measure the concentration of CFC-11 in both the arctic and Antarctic, finding the presence of the gas in each of 50 air samples collected, but incorrectly concluding that CFC's were are not hazardous to the environment. The experiment did however provide the first useful data on the presence of CFC's in the atmosphere.

The damage cased by CFC's discovered by Sherry Rowland and Mario Molina who, after hearing a lecture on the subject of Lovelocks work, embarked on research resulting in the first published paper suggesting the connection in 1974. It turns out that one of CFCs' most attractive features-their unreactivity-has been instrumental in making them one of the most significant pollutants. CFCs' lack of reactivity gives them a lifespan which can exceed 100 years in some cases. This gives them time to diffuse into the upper stratosphere. Here, the sun's ultraviolet radiation is strong enough to break off the chlorine atom, which on its own is a highly reactive free radical. This catalyses the break up of ozone into oxygen by means of a variety of mechanisms, of which the simplest is:

$$Cl\cdot + O_3 \rightarrow ClO\cdot + O_2$$

$$ClO\cdot + O_3 \rightarrow Cl\cdot + 2\ O_2$$

Since the chlorine is regenerated at the end of these reactions, a single Cl atom can destroy many thousands of ozone molecules. Reaction schemes similar to this one (but more complicated) are believed to be the cause of the ozone hole observed over the poles and upper latitudes of the Earth. Decreases in stratospheric ozone may lead to increases in skin cancer.

In 1975, the US state of Oregon enacted the world's first ban of CFCs (legislation introduced by Walter F. Brown). The United States and several European countries banned the use of CFCs in aerosol spray cans in 1978, but continued to use them in refrigeration, foam blowing, and as solvents for cleaning electronic equipment. By 1985, scientists observed a dramatic seasonal depletion of the ozone layer over Antarctica. International attention to CFCs resulted in a meeting of world diplomats in Montreal in 1987. They forged a treaty, the Montreal Protocol, which called for drastic reductions in the production of CFCs. On March 2, 1989, 12 European Community nations agreed to ban the production of all CFCs by the end of the century. In 1990, diplomats met in London and voted to significantly strengthen the Montreal Protocol by calling for a complete elimination of CFCs by the year 2000. By the year 2010 CFCs should be completely eliminated from developing countries as well.

Because the only available CFC gases in countries adhering to the treaty is from recycling, their prices have gone up considerably. A worldwide end to production should also terminate the smuggling of this material, such as from Mexico to the United States.

A number of substitutes for CFCs have been introduced. Hydrochlorofluorocarbons (HCFCs) are much more reactive than CFCs, so a large fraction of the HCFCs emitted break down in the troposphere, and hence are removed before they have a chance to affect the ozone layer. Nevertheless, a significant fraction of the HCFCs do break down in the stratosphere and they have contributed to more chlorine buildup there than originally predicted. Development of non-chlorine based chemical compounds as a substitute for CFCs and HCFCs continues. One such class are the hydrofluorocarbons (HFCs), which contain only hydrogen and fluorine. One of these compounds, HFC-134a, is now used in place of CFC-12 in automobile air conditioners.

There is concern that halons are being broken down in the atmosphere to bromine, which reacts with ozone, leading to depletion of the ozone layer (this is similar to the case of chlorofluorocarbons such as freon). These issues are complicated: the kinds of fires that require halon extinguishers to be put out will typically cause more damage to the ozone layer than the halon itself, not to mention human and property damage. However, fire extinguisher systems must be tested regularly, and these tests may lead to damage. As a result, some regulatory measures have been taken, and halons are being phased out in most of the world.

In the United States, purchase and use of freon gases is regulated by the Environmental Protection Agency, and substantial fines have been levied for their careless venting. Also, licenses, good for life, are required to buy or use these chemicals. The EPA website discusses these rules in great detail, and also lists numerous private companies that are approved to give examinations for these certificates.

There are two kinds of licenses. Obtaining a "Section 609" license to use CFCs to recharge old (pre-1993 model year) car air conditioners is fairly easy and requires only an online multiple choice test offered by several companies. Companies that use unlicensed technicians for CFC recharge operations are subject to a US$15,000 fine per technician by the EPA.

The "Section 608" license, needed to recharge CFC-using stationary and non-automobile mobile units, is also multiple choice but more difficult. A general knowledge test is required, plus separate exams for small size (such as home refrigerator) units, and for high and low pressure systems. These are respectively called Parts I, II, and III. A person who takes and passes all tests receives a "Universal" license; otherwise, one that is endorsed only for the respectively passed Parts. While the general knowledge and Part I exams can be taken online, taking them before a proctor (which has to be done for Parts II and III) lets the applicant pass these tests with lower scores.

Safety

Haloalkanes in copper tubing open to the environment can turn into phosgene gas after coming in contact with extreme heat, such as while brazing or in a fire situation. Other ways that phosgene can be created is by passing the Haloalkane through an internal combustion engine, or by inhaling it through a lit cigarette, cigar or pipe. Phosgene is a substance that was used as a chemical weapon in World War I. Low exposure can cause irritation, but high levels cause fluid to collect in the lungs, possibly resulting in death.

Global Dimming

Global dimming is the gradual reduction in the amount of global direct irradiance at the Earth's surface, observed since the beginning of systematic measurements in 1950s. The effect varies by location, but worldwide it is of the order of a 4% reduction over the three decades from 1960-1990. This trend may have reversed during the past decade. Global dimming creates a cooling effect that may have partially masked the effect of greenhouse gases on global warming.

Cause and Effects

It is currently thought that the effect of global dimming is probably due to the increased presence of aerosol particles in the atmosphere. Aerosol particles and other particulate pollutants absorb solar energy and reflect sunlight back into space. The pollutants can also become nuclei for cloud droplets. It is thought that the water droplets in clouds coalesce around the particles. Increased pollution, resulting in more particulates, creates clouds consisting of a greater number of smaller droplets, which in turn makes them more reflective, therefore bouncing more sunlight back into space.

Clouds intercept both heat from the sun and heat radiated from the Earth. Their effects are complex and vary in time, location and altitude. Usually during the daytime the interception of sunlight predominates, giving a cooling effect; however, at night the re-radiation of heat to the Earth slows the Earth's heat loss.

Research

Early reports of "global dimming" attracted little interest, perhaps because the term itself had not yet been coined. The earliest reports seem to be by M. Budyko: "The effect of solar radiation variations on the climate of the Earth" in 1969, published in Tellus. From the late 1980s onwards, scientists independently began working on solar radiation datasets and discovered declining trends worldwide; Atsumo Ohmura Secular variation of global radiation in Europe in 1989; Vivii Russak in 1990 "Trends of solar radiation, cloudiness and atmospheric transparency during recent decades in Estonia", and Beate Liepert in 1994 "Solar radiation in Germany-Observed trends and an assessment of their causes". Gerry Stanhill who studied these declines worldwide in many papers coined the term "dimming". Dimming exists in sites all over the Former Soviet Union.

Independent research in Israel and the Netherlands in the late 1980s showed an apparent reduction in the amount of sunlight, despite widespread evidence that the climate was actually becoming hotter. The rate of dimming varies around the world but is on average estimated at around 2-3% per decade, with the possibility that the trend reversed in the early 1990s. It is difficult to make a precise measurement, due to the difficulty in accurately calibrating the instruments used, and the problem of spatial coverage. Nonetheless, the effect is almost certainly present.

Note that the effect (2-3%, as above) is due to changes within the Earth's atmosphere; the value of the solar radiation at the top of the atmosphere has not changed by more than a fraction of this amount.

The effect varies greatly over the planet, but estimates of the global average value are:

- 5.3% (9 W/m^2); over 1958-85 (Stanhill and Moreshet, 1992)
- 2%/decade over 1964-93 (Gilgen et al, 1998)
- 2.7%/decade (total 20 W/m^2); up to 2000 (Stanhill and Cohen, 2001)
- 4% over 1961-90 (Liepert 2002)

The largest reductions are found in the northern hemisphere mid-latitudes.

Experiments in the Maldives (comparing the atmosphere over the northern and southern islands) in the 1990s showed that the effect of macroscopic pollutants in the atmosphere at that time (blown south from India) caused about a 10% reduction in sunlight reaching the surface in the area under the pollution cloud-a much greater reduction than expected from the presence of the particles themselves.

Prior to the research being undertaken, predictions were of a 0.5-1% effect from particulate matter; the variation from prediction may be explained by cloud formation with the particles acting as the focus for droplet creation. Clouds are very effective at reflecting light back out into space.

Some climate scientists have theorised that aircraft contrails (also called vapour trails) are implicated in global dimming, but the constant flow of air traffic previously meant that this could not be tested. The near-total shutdown of civil air traffic during the three days following the September 11, 2001 attacks afforded a rare opportunity in which to observe the climate of the United States absent from the effect of contrails. During this period, an increase in diurnal temperature variation of over 1 °C was observed in some parts of the U.S., i.e. aircraft contrails may have been raising nighttime temperatures and/or lowering daytime temperatures by much more than previously thought.

Recent Reversal of the Trend

In 2005 Wild et al. and Pinker et al. found that the "dimming" trend had reversed since about 1990. It is likely that at least some of this change, particularly over Europe, is due to decreases in pollution. Most governments of developed nations have done more to reduce aerosols released into the atmosphere, which helps reduce global dimming, than to reduce CO2 emissions.

The Baseline Surface Radiation Network (BSRN) has been collecting surface measurements. BSRN was started in the early 1990s and updated the archives in this time. Analysis of recent data reveals that the surface of the planet has brightened by about 4% in the past decade. The brightening trend is corroborated by other data, including satellite analyses.

Effects on Rainfall Pattern

Global dimming may have caused large scale changes in weather patterns. Climate models speculatively suggest that this reduction in sunshine at the surface may have led to the failure of the monsoon in sub-Saharan Africa during the 1970s and 1980s, together with the associated famines such as the Sahel drought, caused by Northern hemisphere pollution cooling the Atlantic. Because of this, the Tropical rain belt may not have risen to its northern latitudes, thus causing an absence of seasonal rains. This claim is not universally accepted and is very difficult to test.

It is also concluded that the imbalance between global dimming and global warming at the surface leads to weaker turbulent heat fluxes to the atmosphere. This means globally reduced evaporation and hence precipitation occur in a dimmer and warmer world, which could ultimately lead to a more humid atmosphere in which it rains less.

Relationship to Global Warming

Some scientists now consider that the effects of global dimming have masked the effect of global warming to some extent and that resolving global dimming may therefore lead to increases in predictions of future temperature rise.. It may be suggested that increasing aerosol might be a possible solution to global warming. However, aerosol has negative effects (acid rain) which is why nations have made efforts to reduce it. Also, the aerosol has a very short lifetime (weeks).

The phenomenon underlying global dimming may also have regional effects. While most of the earth has warmed, the regions that are downwind from major sources of air pollution (specifically sulfur dioxide emissions) have generally cooled. This may explain the cooling of the eastern United States relative to the warming western part.

Global Warming

Global warming is the observed increase in the average temperature of the Earth's atmosphere and oceans in recent decades and its projected continuation into the future.

Global average near-surface atmospheric temperature rose 0.6 ± 0.2 °Celsius (1.1 ± 0.4 °Fahrenheit) in the 20th century. The prevailing scientific opinion on climate change is that "most of the warming observed over the last 50 years is attributable to human activities." The main cause of the human-induced component of warming is the increased atmospheric concentration of greenhouse gases (GHGs) such as carbon dioxide (CO_2), which leads to warming of the surface and lower atmosphere by increasing the greenhouse effect. Greenhouse gases are released by activities such as the burning of fossil fuels, land clearing, and agriculture.

Models referenced by the Intergovernmental Panel on Climate Change (IPCC) predict that global temperatures may increase by 1.4 to 5.8 °C (2.5 to 10.5 °F) between 1990 and 2100. The uncertainty in this range results from both the difficulty of predicting the volume of future greenhouse gas emissions and uncertainty about climate sensitivity.

An increase in global temperatures can in turn cause other changes, including a rising sea level and changes in the amount and pattern of precipitation. These changes may increase the frequency and intensity of extreme weather events, such as floods, droughts, heat waves, hurricanes, and tornados. Other consequences include higher or lower agricultural yields, glacier retreat, reduced summer streamflows, species extinctions and increases in the ranges of disease vectors. Warming is expected to affect the number and magnitude of these events; however, it is difficult to connect particular events to global warming. Although most studies focus on the period up to 2100, warming (and sea level rise) is expected to continue past then, since CO_2 has a long average atmospheric lifetime.

Remaining scientific uncertainties include the exact degree of climate change expected in the future, and especially how changes will vary from region to region across the globe. A hotly contested political and public debate has yet to be resolved, regarding whether anything should be done, and what could be cost-effectively done to reduce or reverse future warming, or to deal with the expected consequences. Most national governments have signed and ratified the Kyoto Protocol aimed at combatting global warming.

The term "global warming" is a specific case of the more general term "climate change" (which can also refer to "global cooling," such as occurs during ice ages). In principle, "global warming" is neutral as to the causes, but in common usage, "global warming" generally implies a human influence. However, the UNFCCC uses "climate change" for human-caused change, and "climate variability" for other changes. Some organizations use the term "anthropogenic climate change" for human-induced changes.

History of Warming

Two millennia of mean surface temperatures according to different reconstructions, each smoothed on a decadal scale. The unsmoothed, annual value for 2004 is also plotted for reference.

Relative to the period 1860-1900, global temperatures on both land and sea have increased by 0.75 °C (1.4 °F), according to the instrumental temperature record. Since 1979, land temperatures have increased about twice as fast as ocean temperatures (0.25 °C/decade against 0.13 °C/decade (Smith, 2005). Temperatures in the lower troposphere have increased between 0.12 and 0.22 °C per decade since 1979, according to satellite temperature measurements. Over the one or two thousand years before 1850, world temperature is believed to have been relatively stable, with possibly regional fluctuations such as the Medieval Warm Period or the Little Ice Age.

Based on estimates by NASA's Goddard Institute for Space Studies, 2005 was the warmest year since reliable, widespread instrumental measurements became available in the late 1800s. exceeding the previous record set in 1998 by a few hundredths of a degree. Estimates prepared by the World Meteorological Organization and the UK Climatic Research Unit concluded that 2005 was still only the second warmest year, behind 1998.

Depending on the time frame, a number of temperature records are available based on different data sets. The longest perspective is available from various proxy records for recent millennia; see temperature record of the past 1000 years for a discussion of these records and their differences. An approximately global instrumental record of temperature near the earth's surface begins in about 1860. Global observations of the atmosphere well above the earth's surface using data from radiosondes began shortly after World War II. Satellite temperature measurements of the tropospheric temperature date from 1979. The attribution of recent climate change is clearest for the most recent period of the last 50 years, for which the most detailed data are available.

Causes

Carbon dioxide during the last 400,000 years and the rapid rise since the Industrial Revolution; changes in the Earth's orbit around the Sun, known as Milankovitch cycles, are believed to be the pacemaker of the 100,000 year ice age cycle.

The climate system varies both through natural, "internal" processes as well as in response to variations in external "forcing" from both human and non-human causes, including solar activity, volcanic emissions, variations in the earth's orbit ("orbital forcing") and greenhouse gases. Climatologists agree that the earth has warmed recently. The detailed causes of this change remain an active field of research, but the scientific consensus identifies greenhouse gases as the primary cause of the recent warming. Outside of the scientific community, however, this conclusion can be controversial.

Adding carbon dioxide (CO_2) or methane (CH_4) to Earth's atmosphere, with no other changes, will make the planet's surface warmer; greenhouse gases create a natural greenhouse effect without which temperatures on Earth would be an estimated 30 °C (54 °F) lower, and the Earth uninhabitable. It is therefore not correct to say that there is a debate between those who "believe in" and "oppose" the theory that adding carbon dioxide or methane to the Earth's atmosphere will, absent any mitigating actions or effects, result in warmer surface temperatures on Earth. Rather, the debate is about what the net effect of the addition of carbon dioxide and methane will be, when allowing for compounding or mitigating factors.

One example of an important feedback process is ice-albedo feedback. The increased CO_2 in the atmosphere warms the Earth's surface and leads to melting of ice near the poles. As the ice melts, land or open water takes its place. Both land and open water are on average less reflective than ice, and thus absorb more solar radiation. This causes more warming, which in turn causes more melting, and the cycle continues.

Due to the thermal inertia of the earth's oceans and slow responses of other indirect effects, the Earth's current climate is not in equilibrium with the forcing imposed by increased greenhouse gases. Climate commitment studies indicate that, even if greenhouse gases were stabilized at present day levels, a further warming of perhaps 0.5 °C to 1.0 °C (0.9-1.8 °F) would still occur.

Greenhouse Gases in the Atmosphere

Greenhouse gases are transparent to shortwave radiation from the sun, the main source of heat on the Earth. However, they absorb some of the longer infrared radiation emitted by the Earth, thereby reducing radiational cooling and hence raising the temperature of the Earth. How much they warm the world by is shown in their global warming potential. The measure of the response to increased GHGs, and other anthropogenic and natural climate forcings is climate sensitivity. It is found by observational and model studies.

This sensitivity is usually expressed in terms of the temperature response expected from a doubling of CO_2 in the atmosphere. The current literature estimates sensitivity in the range 1.5-4.5 °C (2.7-8.1 °F).

The atmospheric concentrations of carbon dioxide and methane have increased by 31% and 149% respectively above pre-industrial levels since 1750. This is considerably higher than at any time during the last 650,000 years, the period for which reliable data has been extracted from ice cores. From less direct geological evidence it is believed that carbon dioxide values this high were last attained 40 million years ago. About three-quarters of the anthropogenic (man-made) emissions of carbon dioxide to the atmosphere during the past 20 years are due to fossil fuel burning. The rest of the anthropogenic emissions are predominantly due to land-use change, especially deforestation.

The longest continuous instrumental measurement of carbon dioxide mixing ratios began in 1958 at Mauna Loa. Since then, the annually averaged value has increased monotonically by approximately 21% from the initial reading of 315 ppmv, as shown by the Keeling curve, to over 380 ppmv in 2006. The monthly CO_2 measurements display small seasonal oscillations in an overall yearly uptrend; each year's maximum is reached during the northern hemisphere's late spring and declines during the northern hemisphere growing season as plants remove some CO_2 from the atmosphere.

Methane, the primary constituent of natural gas, enters the atmosphere both from biological production and leaks from natural gas pipelines and other infrastructure. Some biological sources are natural, such as termites or forests, but others have been increased or created by agricultural activities such as the cultivation of rice paddies. Recent evidence indicates that methane concentrations have begun to stabilize, perhaps due to reductions in leakage from fuel transmission and storage facilities.

Future carbon dioxide levels are expected to continue rising due to ongoing fossil fuel usage. The rate of rise will depend on uncertain economic, sociological, technological, and natural developments. The IPCC Special Report on Emissions Scenarios gives a wide range of future carbon dioxide scenarios, ranging from 541 to 970 parts per million by the year 2100. Fossil fuel reserves are sufficient to reach this level and continue emissions past 2100, if coal and tar sands are extensively used.

Carbon sink ecosystems (forests and oceans) are being degraded by pollutants. Degradation of major carbon sinks results in higher atmospheric carbon dioxide levels.

Globally, the majority of anthropogenic greenhouse gas emissions arise from fuel combustion. The remainder is accounted for largely by "fugitive fuel" (fuel consumed in the production and transport of fuel), emissions from industrial processes (excluding fuel combustion), and agriculture: these contributed 5.8%, 5.2% and 3.3% respectively in 1990. Current figures are broadly comparable. Around 17% of emissions are accounted for by the combustion of fuel for the generation of electricity. A small percentage of emissions come from natural and anthropogenic biological sources, with approximately 6.3% derived from agriculturally produced methane and nitrous oxide.

Positive feedback effects, such as the expected release of methane from the melting of permafrost peat bogs in Siberia (possibly up to 70,000 million tonnes), may lead to significant additional sources of greenhouse gas emissions. Note that the anthropogenic emissions of other pollutants-notably sulfate aerosols-exert a cooling effect; this partially accounts for the plateau/cooling seen in the temperature record in the middle of the twentieth century, though this may also be due to intervening natural cycles.

Other Hypotheses

The extent of the scientific consensus on global warming-that "most of the observed warming over the last 50 years is likely to have been attributable to human activities"-has been investigated: In the journal Science in December 2004, Dr Naomi Oreskes published a study of the abstracts of the 928 refereed scientific articles in the ISI citation database identified with the keywords "global climate change" and published from 1993-2003. This study concluded that 75% of the 928 articles either explicitly or implicitly accepted the consensus view-the remainder of the articles covered methods or paleoclimate and did not take any stance on recent climate change. The study did not report how many of the 928 abstracts explicitly accepted the hypothesis of human-induced warming, but none of the 928 articles surveyed accepted any other hypothesis. Contrasting with the consensus view, other hypotheses have been proposed to explain all or most of the observed increase in global temperatures. Some of these hypotheses include:

- The warming is within the range of natural variation.
- The warming is a consequence of coming out of a prior cool period, namely the Little Ice Age.
- The warming is primarily a result of variances in solar irradiance, possibly via modulation of cloud cover. It is similar in concept to the operating principles of the Wilson cloud chamber, but on a global scale where earth's atmosphere acts as the cloud chamber and the cosmic rays catalyze the production of cloud condensation nuclei.
- The observed warming actually reflects the Urban Heat Island, as most readings are done in heavily populated areas which are expanding with growing population.

The Solar Variation Theory

Modeling studies reported in the IPCC Third Assessment Report (TAR) did not find that changes in solar forcing were needed in order to explain the climate record for the last four or five decades. These studies found that volcanic and solar forcings may account for half of the temperature variations prior to 1950, but the net effect of such natural forcings has been roughly neutral since then. In particular, the change in climate forcing from greenhouse gases since 1750 was estimated to be eight times larger than the change in forcing due to increasing solar activity over the same period.

Since the TAR, some studies (Lean et al., 2002, Wang et al., 2005) have suggested that changes in irradiance since pre-industrial times are less by a factor of 3 to 4 than in the reconstructions used in the TAR (e.g. Hoyt and Schatten, 1993, Lean, 2000.). Other researchers (e.g. Stott et al. 2003) believe that the effect of solar forcing is being underestimated and propose that solar forcing accounts for 16% or 36% of recent greenhouse warming. Others (e.g. Marsh and Svensmark 2000) have proposed that feedback from clouds or other processes enhance the direct effect of solar variation, which if true would also suggest that the effect of solar variability was being underestimated. In general the level of scientific understanding of the contribution of variations in solar irradiance to historical climate changes is "very low".

The present level of solar activity is historically high. Solanki et al. (2004) suggest that solar activity for the last 60 to 70 years may be at its highest level in 8,000 years; Muscheler et al. disagree, suggesting that other comparably high levels of activity have occurred several times in the last few thousand years. Solanki concluded based on their analysis that there is a 92% probability that solar activity will decrease over the next 50 years. In addition, researchers at Duke University (2005) have found that 10-30% of the warming over the last two decades may be due to increased solar output. In a review of existing literature, Foukal et al. (2006) determined both that the variations in solar output were too small to have contributed appreciably to global warming since the mid-1970s and that there was no evidence of a net increase in brightness during this period.

Attributed and Expected Effects

Global glacial mass balance in the last fifty years, reported to the WGMS and the NSIDC. The increased downward trend in the late 1980s is symptomatic of the increased rate and number of retreating glaciers. Some effects on both the natural environment and human life are already being attributed at least in part to global warming. Glacier retreat, ice shelf disruption such as of the Larsen Ice Shelf, sea level rise, changes in rainfall patterns, increased intensity and frequency of hurricanes and extreme weather events, are being attributed at least in part to global warming. While changes are expected for overall patterns, intensity, and frequencies, it is difficult or impossible to attribute specific events (such as Hurricane Katrina) to global warming.

Some anticipated effects include sea level rise of up to 14 metres, repercussions to agriculture, possible slowing of the thermohaline circulation, reductions in the ozone layer, increased intensity and frequency of hurricanes and extreme weather events, lowering of ocean pH, the spread of diseases such as malaria and dengue fever, and mass extinction events.

The extent and probability of these consequences is a matter of nsiderable uncertainty. A summary of probable effects and rec t understanding can be found in the report of the IPCC W ing Group II.

Mitigation

The Energy Information Administration predicts world energy and fossil fuel usage will rise in the next decades. The consensus among climate scientists that global temperatures will continue to increase has led nations, states, corporations and individuals to try to reduce its effects. Their actions and proposals fall into five categories:

1. Reduction of energy use
2. Shifting from carbon-based fossil fuels to alternative energy sources
3. Carbon capture and storage
4. Carbon sequestration
5. Planetary engineering to cool the earth, including screening out sunlight and increasing the reflectivity of the earth.

Mitigation Strategies

Strategies for mitigation of global warming include development of new technologies; carbon offsets; renewable energy such as biodiesel, wind power, and solar power; nuclear power; electric or hybrid automobiles; fuel cells; energy conservation; carbon taxes; improving natural carbon dioxide sinks; increased use of sulfate aerosols, which exhibit a cooling effect on the Earth; population control; and carbon capture and storage. Many environmental groups encourage individual action against global warming, often aimed at the consumer, and there has been business action on climate change. The world's primary international agreement on combating climate change is the Kyoto Protocol. The Kyoto Protocol is an amendment to the United Nations Framework Convention on Climate Change (UNFCCC). Countries that ratify this protocol commit to reduce their emissions of carbon dioxide and five other greenhouse gases, or engage in emissions trading if they maintain or increase emissions of these gases.

Climate Models

Scientists have studied global warming with computer models of the climate. Before the scientific community accepts a climate model, it has to be validated against observed climate variations. As of 2006, models with sufficiently high resolution are able to successfully simulate summer/winter differences, the North Atlantic Oscillation, and El Niño. All validated current models predict that the net effect of adding greenhouse gases will be a warmer climate in the future.

However, the amount of predicted warming varies by model, and there still remains a considerable range of climate sensitivity predicted by the models which survive these tests; one of the most important sources of this uncertainty is believed to be different ways of handling clouds. Part of the technical summary of the IPCC TAR includes a recognition of the need to quantify this uncertainty: "In climate research and modeling, we should recognize that we are dealing with a coupled non-linear system, and therefore that the prediction of a specific future climate is not possible. Rather the focus must be on the probability distribution of the system's possible future states by the generation of ensembles of model solutions."

As noted above, climate models have been used by the IPCC to anticipate a warming of 1.4°C to 5.8°C (2.5 °F-10.4 °F) between 1990 and 2100. They have also been used to help investigate the causes of recent climate change by comparing the observed changes to those that the models predict from various natural and human derived forcing factors. In addition to having their own characteristic climate sensitivity, models have also been used to derive independent assessments of climate sensitivity. Climate models can produce a good match to observations of global temperature changes over the last century. These models do not unambiguously attribute the warming that occurred from approximately 1910 to 1945 to either natural variation or human effects; however, they suggest that the warming since 1975 is dominated by man-made greenhouse gas emissions. Adding simulation of the carbon cycle to the models generally shows a positive feedback, though this response is uncertain (under the A2 SRES scenario, responses vary between an extra 20 and 200 ppm of CO_2). Some observational studies also show a positive feedback.

Uncertainties in the representation of clouds are a dominant source of uncertainty in existing models, despite clear progress in modeling of clouds. There is also an ongoing discussion as to whether climate models are neglecting important indirect and feedback effects of solar variability. Further, all such models are limited by available computational power, so that they may overlook changes related to small-scale processes and weather (e.g. storm systems, hurricanes). However, despite these and other limitations, the IPCC considered climate models "to be suitable tools to provide useful projections of future climates".

In December, 2005 Bellouin et al. suggested in Nature that the reflectivity effect of airborne pollutants was about double that previously expected, and that therefore some global warming was being masked. If supported by further studies, this would imply that existing models under-predict future global warming.

Ocean Acidification

Increased atmospheric carbon dioxide increases the amount of CO_2 dissolved in the oceans. Unfortunately, carbon dioxide gas dissolved in the ocean reacts with water to form carbonic acid resulting in ocean acidification. Since biosystems are adapted to a narrow range of pH this is a serious concern directly driven by increased atmospheric CO_2 and not global warming.

Relationship to Ozone Depletion

Although they are often interlinked in the mass media, the connection between global warming and ozone depletion is not strong. There are five areas of linkage:

- The same carbon dioxide radiative forcing that produces near-surface global warming is expected (perhaps somewhat surprisingly) to cool the stratosphere. This, in turn, would lead to a relative increase in ozone depletion and the frequency of ozone holes.

Radiative Forcing from Various Greenhouse Gases and Other Sources

- Conversely, ozone depletion represents a radiative forcing of the climate system. There are two opposed effects:

(1) reduced ozone allows more solar radiation to penetrate, thus warming the troposphere instead of the stratosphere.

(2) The resulting colder stratosphere emits less long-wave radiation down to the troposphere, thus having a cooling effect. Overall, the cooling dominates: the IPCC concludes that observed stratospheric O_3 losses over the past two decades have caused a negative forcing of the surface-troposphere system of about-0.15 ± 0.10 W/m^2.

- One of the strongest predictions of the greenhouse effect theory is that the stratosphere will cool. Although this cooling has been observed, it is not trivial to separate the effects of changes in the concentration of greenhouse gases and ozone depletion since both will lead to cooling. However, this can be done by numerical stratospheric modeling. Results from the NOAA Geophysical Fluid Dynamics Laboratory show that above 20 km, the greenhouse gases dominate the cooling.
- Ozone depleting chemicals are also greenhouse gases, representing 0.34 ± 0.03 W/m^2, or about 14% of the total radiative forcing from well-mixed greenhouse gases.
- Decreased ozone leads to an increase in ultraviolet levels. Ultraviolet radiation may be responsible for the death of ocean algae, which operate as a carbon dioxide sink in the ocean. Increased UV, therefore, may lead to a decrease in carbon dioxide uptake, thereby raising global carbon dioxide levels.

Relationship to Global Dimming

Some scientists now consider that the effects of global dimming (the reduction in sunlight reaching the surface of the planet, possibly due to aerosols) may have masked some of the effect of global warming. If this is so, the indirect aerosol effect is stronger than previously believed, which would imply that the climate sensitivity to greenhouse gases is also stronger. Concerns about the effect of aerosol on the global climate were first researched as part of concerns over global cooling in the 1970s.

Pre-human Global Warming

The Earth has experienced natural global warming and cooling many times in the past, and can offer useful insights into present processes. It is thought by some geologists that a rapid buildup of greenhouse gases caused the Earth to experience global warming in the early Jurassic period, with average temperatures rising by 5 °C (9.0 °F). Research by the Open University published in Geology (32: 157-160, 2004) indicates that this caused the rate of rock weathering to increase by 400%. As such weathering locks away carbon in calcite and dolomite, carbon dioxide levels dropped back to normal over roughly the next 150,000 years.

Sudden releases of methane from clathrate compounds (the Clathrate Gun Hypothesis) have been hypothesized as a cause for other past global warming events, including the Permian-Triassic extinction event and the Paleocene-Eocene Thermal Maximum. However, warming at the end of the last glacial period is thought not to be due to methane release. Instead, natural variations in the Earth's orbit (Milankovitch cycles) are believed to have triggered the retreat of ice sheets by changing the amount of solar radiation received at high latitude and led to deglaciation.

The greenhouse effect is also invoked to explain how the Earth made it out of the Snowball Earth period 600 million years ago. During this period all silicate rocks were covered by ice, thereby preventing them from combining with atmospheric carbon dioxide. The atmospheric carbon dioxide level gradually increased until it reached a level that could have been as much as 350 times the current level. At this point temperatures were raised enough to melt the ice, even though the reflective ice surfaces had been reflecting most sunlight back into space. Increased amounts of rainfall would quickly wash the carbon dioxide out of the atmosphere, and thick layers of abiotic carbonate sediment have been found on top of the glacial rocks from this period.

Using paleoclimate data for the last 500 million years Veizer et al. (2000, Nature 408, pp. 698-701) concluded that long-term temperature variations are only weakly related to carbon dioxide variations. Most paleoclimatologists believe this is because other factors, such as continental drift and mountain building have larger effects in determining very long term climate. However, Shaviv and Veizer, proposed that the biggest long-term influence on temperature is actually the solar system's motion around the galaxy, and the ways in which this influences the atmosphere by altering the flux of cosmic rays received by the Earth. Afterwards, they argued that over geologic times a change in carbon dioxide concentrations comparable to doubling pre-industrial levels, only results in about 0.75 °C (1.3 °F) warming rather than the usual 1.5-4.5 °C (2.7-8.1 °F) reported by climate models. They acknowledge (Shaviv and Veizer 2004) however that this conclusion may only be valid on multi-million year time scales when glacial and geological feedback have had a chance to establish themselves. Rahmstorf et al. 2004 argue that Shaviv and Veizer arbitrarily tuned their data, and that their conclusions are unreliable.

Pre-industrial Global Warming

Paleoclimatologist William Ruddiman has argued that human influence on the global climate began around 8,000 years ago with the start of forest clearing to provide land for agriculture and 5,000 years ago with the start of Asian rice irrigation. He contends that forest clearing explains the rise in carbon dioxide levels in the current interglacial that started 8,000 years ago, contrasting with the decline in carbon dioxide levels seen in the previous three interglacials. He further contends that the spread of rice irrigation explains the breakdown in the last 5,000 years of the correlation between the Northern Hemisphere solar radiation and global methane levels, which had been maintained over at least the last 11 22,000-year cycles. Ruddiman argues that without these effects, the Earth would be nearly 2 °C cooler and "well on the way" to a new ice age. Ruddiman's interpretation of the historical record, with respect to the methane data, has been disputed.

Deforestation

While forests are being cleared permanently in nearly all parts of the world, the issue of deforestation generally focuses on the loss of tropical forests. Tropical forests, including rain forests, are confined to equatorial areas, especially Latin

America, Africa and Southern and Southeastern Asia. Tropical forests currently cover slightly over 5% of the earth's land area, but are rapidly being permanently altered-it is reliably estimated that we have already lost or damaged over 50% of the world's tropical forest resources. The problem is most severe in the less developed nations-Haiti and the Philippines have already lost more than 95% of their tropical forests. Why is this a problem? After all, trees can be replanted. And any way, how does this affect us in the more developed nations? In fact, deforestation is a global environmental issue, one which affects all of us, whether we live in an area being deforested or in an area in which people are trying to protect their forests, especially tropical and rain forests. As with so many other environmental issues, this one too often pits less developed nations on one side of the issue and developed nations on the other. To understand why deforestation is taking place, at an alarming rate, we need first to identify why the forests are being cut. Most of these reasons can be grouped together as the perceived and real different needs for forested land.

Perceived and Real Different Needs for Forested Land

One might think that the primary reason for eliminating tropical forests is for the economic value of the timber being removed. In some cases, this appears to be the rationale. It certainly was a large part of the reason for the near elimination of teak forests in Thailand, and certainly, it is a part of the reason for most deforestation. However, in too many cases, the primary reason eliminating forest land is to convert it to cropland and grazing land, economically more important in the short term for burgeoning populations and poverty in less developed nations. Subsistence farmers feel they have no alternative but to cut the forests in order to grow enough food for their families to survive.

But what are the long-term costs? These can be put into five major, sometimes related groups:

Loss of Forest Commodities

In addition to the hardwood timber contained therein, tropical forests supply people of the world with hundreds of food products (coffee, spices, chocolate, tropical fruits and nuts, etc.), as well as such important non-edible commodities as active ingredients for prescription drugs, natural rubber, oils, and dyes, that cannot be produced elsewhere, and which could be harvested sustainably if the forests are not destroyed. It has been estimated that sustainable harvesting of these commodities could yield two-to three-times the economic value as timber production and cattle grazing, over a 50-year period. Here again, however, we see how the battle between short-term survivability takes precedent over long-term sustainability.

Soil Erosion

Soil erosion is perhaps the most visible of the effects of deforestation. While soil erosion is a problem in much of the world, it is even more severe in tropical regions, where rain fall is highest, and previously ungrazed land is now used for large-scale grazing by large animals. It is too often ironic that forested land is converted to cropland and grazing land, only to have it become an unproductive wasteland due to extensive soil erosion. If the problem stopped there, it would be serious enough, but the subsistence farmers and ranchers who cut the tropical forests and do not manage their soil resources properly, too often just pack up their families and move to an adjacent area, starting the process all over again, depleting even more of the forest resources, and turning even more valuable land into land much less valuable.

A related effect of deforestation is water pollution. Click here for a discussion of water pollution related to deforestation.

Water Pollution

Almost as visible as soil erosion in denuded tropical forests is the related and nearby water pollution. Soil erosion in deforested areas causes there to be a significant influx of soil (sediment) into the surface waters, increasing the turbidity, or ability of water to transmit light. Many animals and plants can tolerate only narrow ranges of turbidity. Those which cannot tolerate the increased turbidity will no longer exist in this particular ecosystem. As one or more plants or animals dies out in an ecosystem, all other life forms in the food chains containing the now-nonexistent plant or animal are affected.

Increased sediment load in rivers will also affect the coastal lands at the mouths of the rivers. Much of the sediment transported by the rivers to the coastal zone is deposited when the river slows and when the fresh water in the river system mixes with the salt water of the ocean. If the sediment load of a river is significantly increased, so too will be the amount of sediment deposited at the mouth of the river. In turn this can affect how the land is used and the numbers and types of plants and animals in that ecosystem.

Habitat Destruction

Tropical deforestation is the leading factor in the destruction of habitats of many plants and animals. Habitat destruction, in turn, is the leading cause of extinctions of both plants and animals. As trees and other plants are cut, the habitats of many animals are literally destroyed. Many plants and animals can only live in tropical forests. When the tropical forests are gone, so will be the plants and animals which once lived there.

Climate Change

Deforestation not only leads to the loss of habitats, but contributes to climate change by adding carbon dioxide to the atmosphere. Growing trees absorb carbon dioxide from the air in the process of photosynthesis, thus buffering, to some extent, the amount of carbon dioxide in the air. When cut, trees can no longer use atmospheric carbon dioxide. Even more carbon dioxide is released when the trees are burned or left to decay. Deforestation is second only to the burning of fossil fuels as a source of increased atmospheric carbon dioxide. Millions of acres of tropical forests are being destroyed to create pastures and farm land, and to supply forest products and wood fuel to increasing human population demands. An increase in carbon dioxide in the atmosphere can enhance the greenhouse effect and global warming.

Saving Tropical Forests

It should be clear by now that tropical forests need to be saved, to the extent possible. It is also clear what practices need to be effected now to begin the process of saving this precious global resource. It is not necessary to completely eliminate the cutting of tropical forests in order to save them. Many of the problems noted in the preceding sections begin when large areas of forests are cleared, and nothing is replanted and nothing is done to protect the soil. On the other hand, when small areas of forests are clear cut, and reseeded, then soil fertility can be restored and many of the problems noted in the preceding sections can be eliminated or at least significantly reduced. However, the root cause of deforestation problems lie in the burgeoning human population growth in areas of the world where the tropical forests exist. Most of the tropical nations of South America and Africa have among the highest population growth rates and lowest per capita incomes.

Pressures from population growth-the need for space and food especially-cause many nations to allow its peoples to engage in practices which while perhaps effective in the short term, are clearly detrimental or devastational in the long term. Reducing population growth rates will aid in the addressing of many of the problems associated with deforestation.

El Nino-A Consequence of Global Warming?

As stated earlier, there is a relationship between the climate system and other Earth systems. Southeast Trade Winds generally blow warm surface equatorial water off shore causing colder deeper ocean water to rise to the surface. This upwelling brings nutrient rich water to the surface which encourages phytoplankton and zooplankton growth. This typical system of equatorial currents also is responsible for a generally arid climate in western South America and generally humid or even monsoonal climate in the western Pacific (Australia, New Zealand, Malaysia, Philippines, etc.)

Periodically, the Southeastern Trade Winds weaken and the warm Equatorial Current moves toward the South American coast. This event is known as the El Nino-Southern Oscillation (ENSUE). This event was first recognized by fishermen off the coast of Ecuador and Peru around the Christmas season thereby giving it the name of "The Child" referring to the Christ Child. El Nino events are cyclic in nature, generally occurring every 4 to 7 years. Recent El Nino events have occurred in 1972-73, 1976-77, 1982-83, and 1997-98.

Effects of El Nino Events

El Nino events result in extensive shifts in sea water temperatures, which in turn cause large scale changes in weather patterns all over the planet. Climate changes associated with El Nino events are storms, droughts, and floods all over the world. The normally arid west coast of South America becomes very wet, and the normally wet western Pacific becomes very arid.

The El Nino of 1982-83, called the "Super El Nino", caused droughts in Africa, Australia, and Indonesia; flooding in Tahiti, California, Ecuador, and Peru; and larger than normal waves in California and on the east coast of the U.S., causing severe beach erosion. The El Nino event which occurred in 1997-98, is now considered to have had an even more detrimental effect on the Earth than the 1982-82 event.

Causes for El Nino Events

The causes for the onset of El Nino events are not yet scientifically established, but apparently are not related to actions that can be directly attributed to human activity. So, why are we including a discussion of the effects of El Nino events in this unit on global change? Even though there is not yet an established relationship between human actions and the development of El Nino events, scientists are now conducting studies to establish if there is an increased frequency and intensity of El Nino events which might be related to global warming.

WATER POLLUTION

Water pollution is any physical, chemical or biological alteration of water that is harmful to living organisms. A more human-centered definition describes pollution as "any contamination of water that makes it unsuitable for its 'highest' use". For example, silt or mud in water would not affect its use for boating or fishing, but the same silt in the same water could prevent humans from drinking or washing clothes in it. There are many types of water pollution; some have a natural origin, but most are related to human activities.

Water becomes polluted easily because it is a nearly "universal" solvent. One of the alchemists' goals was to find a "universal solvent"-a liquid into which everything could be dissolved. They were unaware that they already possessed what they were looking for-almost everything will dissolve in water, at least to some extent.

This makes water an excellent cleaner, and a convenient way to dispose of waste. These initially positive uses, in turn, often lead to water pollution. There are many sources of water pollution, but a common way to categorize them is to consider point sources and non-point sources. Point sources are specific locations that are easy to identify. Less specific locations, more generalized and harder to find, are called non-point sources, and are often difficult to define and to regulate. Whatever its cause, water pollution does have many possible solutions.

Pollution in streams and lakes is often easy to see or smell, and frequently the solutions are obvious. Groundwater, though hidden below the surface, can also become polluted; and because it is harder to see and easier to forget, this pollution is more difficult to detect and to clean up.

Water pollution is a large set of adverse effects upon water bodies (lakes, rivers, oceans, groundwater) caused by human activities. Although natural phenomena such as volcanoes, storms, earthquakes etc. also cause major changes in water quality and the ecological status of water, these are not deemed to be pollution. Water pollution has many causes and characteristics. Increases in nutrient loading may lead to eutrophication.

Organic wastes such as sewage and farm waste impose high oxygen demands on the receiving water leading to oxygen depletion with potentially severe impacts on the whole eco-system. Industries discharge a variety of pollutants in their wastewater including heavy metals, organic toxins, oils, nutrients, and solids. Discharges can also have thermal effects, especially those from power stations, and these too reduce the available oxygen.

Silt-bearing runoff from many activities including construction sites, deforestation and agriculture can inhibit the penetration of sunlight through the water column, restricting photosynthesis and causing blanketing of the lake or river bed, in turn damaging ecological systems. Pollutants in water include a wide spectrum of chemicals, pathogens, and physical chemistry or sensory changes, though the main cause of pollution in both lakes and rivers are animal wastes and fertilizer. Many of the chemical substances are toxic or even carcinogenic. Pathogens can obviously produce waterborne diseases in either human or animal hosts. Alteration of water's physical chemistry include acidity, conductivity, temperature, and eutrophication. Eutrophication is the fertilisation of surface water by nutrients that were previously scarce. Even many of the municipal water supplies in developed countries can present health risks.

Water pollution is a serious problem in the global context. It has been suggested that it is the leading worldwide cause of deaths and diseases, and that it accounts for the deaths of more than 14,000 people daily.

Sources of Water Pollution

Principal sources of water pollution are:

- industrial discharge of chemical wastes and byproducts
- discharge of poorly-treated or untreated sewage
- surface runoff containing pesticides
- slash and burn farming practice, which is often an element within shifting cultivation agricultural systems
- surface runoff containing spilled petroleum products
- surface runoff from construction sites, farms, or paved and other impervious surfaces e.g. silt
- discharge of contaminated and/or heated water used for industrial processes
- acid rain caused by industrial discharge of sulfur dioxide (by burning high-sulfur fossil fuels)
- excess nutrients added by runoff containing detergents or fertilizers
- underground storage tank leakage, leading to soil contamination, thence aquifer contamination.

Contaminants

Contaminants may include organic and inorganic substances.

Some organic water pollutants are:

- insecticides and herbicides, a huge range of organohalide and other chemicals
- bacteria, often is from sewage or livestock operations;
- food processing waste, including pathogens
- tree and brush debris from logging operations
- VOCs (Volatile Organic Compounds, industrial solvents) from improper storage

Some inorganic water pollutants include:

- heavy metals including acid mine drainage
- acidity caused by industrial discharges (especially sulfur dioxide from power plants)
- chemical waste as industrial by products
- fertilizers, in runoff from agriculture including nitrates and phosphates
- silt in surface runoff from construction sites, logging, slash and burn practices or land clearing sites

Transport and Chemical Reactions of Water Pollutants

Most water pollutants are eventually carried by the rivers into the oceans. In some areas of the world the influence can be traced hundred miles from the mouth by studies using hydrology transport models. Advanced computer models such as SWMM or the DSSAM Model have been used in many locations worldwide to examine the fate of pollutants in aquatic systems. Indicator filter feeding species such as copepods have also been used to study pollutant fates in the New York Bight, for example. The highest toxin loads are not directly at the mouth of the Hudson River, but 100 kilometers south, since several days are required for incorporation into planktonic tissue. The Hudson discharge flows south along the coast due to coriolis force. Further south then are areas of oxygen depletion, caused by chemicals using up oxygen and by algae blooms, caused by excess nutrients from algal cell death and decomposition. Fish and shellfish kills have been reported, because toxins climb the foodchain after small fish consume copepods, then large fish eat smaller fish, etc. Each step up the food chain concentrates certain toxins like heavy metals and DDT by approximately a factor of ten.

For several years ocean researcher Charles Moore has been investigating a concentration of floating plastic debris in the Pacific Ocean. His study indicates that ocean currents have added to the mass until it is now about the size of Texas. Many of these long-lasting pieces wind up in the stomachs of marine birds and animals.

Many chemicals undergo reactive decay or change especially over long periods of time in groundwater reservoirs. A noteworthy class of such chemicals are the chlorinated hydrocarbons such as trichloroethylene (used in industrial metal degreasing) and tetrachloroethylene used in the dry cleaning industry. Both of these chemicals, which are carcinogens themselves, undergo partial decomposition reactions leading to new hazardous chemicals.

Groundwater pollution is much more difficult to abate than surface pollution because groundwater can move great distances through unseen aquifers. Non-porous aquifers such as clays partially purify water of bacteria by simple filtration (adsorption and absorption), dilution, and, in some cases, chemical reactions and biological activity: however, in some cases, the pollutants merely transform to soil contaminants. Groundwater that moves through cracks and caverns is not filtered and can be transported as easily as surface water. In fact this can be aggravated by the human tendency to use natural sinkholes as dumps in areas of Karst topography.

There are a variety of secondary effects stemming not from the original pollutant, but a derivative condition. Some of these secondary impacts are:

- Silt bearing surface runoff from can inhibit the penetration of sunlight through the water column, hampering Photosynthesis in aquatic plants.

- Thermal pollution can induce fish kills and invasion by new thermophyllic species

Regulatory Framework

In the UK there are common law rights to protect the passage of water across land unfettered in either quality of quantity. Criminal laws dating back to the 16th century excercised some control over water pollution but it was not until the River (Prevention of pollution) Acts 1951-1961 were enacted that any systematic control over water pollution was established. These laws were strengthened and extended in the Control of Pollution Act 1984 which has since been updated and modified by a series of further acts.

In the USA, concern over water pollution resulted in the enactment of state anti-pollution laws in the latter half of the 19th century, and federal legislation enacted in 1899. The Refuse Act of the federal Rivers and Harbors Act of 1899 prohibits the disposal of any refuse matter from into either the nation's navigable rivers, lakes, streams, and other navigable bodies of water, or any tributary to such waters, unless one has first obtained a permit. The Water Pollution Control Act, passed in 1948, gave authority to the Surgeon General to reduce water pollution.

Growing public awareness and concern for controlling water pollution led to enactment of the Federal Water Pollution Control Act Amendments of 1972. As amended in 1977, this law became commonly known as the Clean Water Act. The Act established the basic mechanisms for regulating contaminant discharge. It established the authority for EPA to implement wastewater standards for industry. The Clean Water Act also continued requirements to set water quality standards for all contaminants in surface waters. Further amplification of the Act continued including the enactment of the Great Lakes Legacy Act of 2002 (Public Law 107-303, November 27, 2002).

EUTROPHICATION

Eutrophication is caused by the decrease of an ecosystem with chemical nutrients, typically compounds containing nitrogen or phosphorus. It may occur on land or in the water.

Eutrophication is frequently a result of nutrient pollution such as the release of sewage effluent into natural waters (rivers or coasts) although it may occur naturally in situations where nutrients accumulate (e.g. depositional environments) or where they flow into systems on an ephemeral basis (e.g. intermittent upwelling in coastal systems). Eutrophication generally promotes excessive plant growth and decay, favors certain weedy species over others, and is likely to cause severe reductions in water quality. In aquatic environments, enhanced growth of choking aquatic vegetation or phytoplankton (that is, an algal bloom) disrupts normal functioning of the ecosystem, causing a variety of problems. Human society is impacted as well: eutrophication decreases the resource value of rivers, lakes, and estuaries such that recreation, fishing, hunting, and aesthetic enjoyment are hindered. Health-related problems can occur where eutrophic conditions interfere with drinking water treatment. Although traditionally thought of as enrichment of aquatic systems by addition of fertilizers into lakes, bays, or other semi-enclosed waters (even slow-moving rivers), terrestrial ecosystems are subject to similarly adverse impacts. Increased content of nitrates in soil frequently leads to undesirable changes in vegetation composition and many plant species are endangered as a result of eutrophication in terrestric ecosystems, e.g. majority of orchid species in Europe. Ecosystems (like some meadows, forests and bogs that are characterized by low nutrient content and species-rich, slowly growing vegetation adapted to lower nutrient levels) are overgrown by faster growing and more competitive species-poor vegetation, like tall grasses, that can take advantage of unnaturally elevated nitrogen level and the area may be changed beyond recognition and vulnerable species may be lost. Eg. species-rich fens are overtaken by reed or reedgrass species, spectacular forest undergrowth affected by run-off from nearby fertilized field is turned into a thick nettle and bramble shrub.

Eutrophication was recognized as a pollution problem in European and North American lakes and reservoirs in the mid-20th century. Since then, it has become more widespread. Surveys showed that 54% of lakes in Asia are eutrophic; in Europe, 53%; in North America, 48%; in South America, 41%; and in Africa, 28%.

Concept of Eutrophication

Eutrophication can be a natural process in lakes, as they fill in through geological time, though other lakes are known to demonstrate the reverse process, becoming less nutrient rich with time. Estuaries also tend to be naturally eutrophic because land-derived nutrients are concentrated where run-off enters the marine environment in a confined channel and mixing of relatively high nutrient fresh water with low nutrient marine water occurs.

Phosphorus is often regarded as the main culprit in cases of eutrophication in lakes subjected to point source pollution from sewage. The concentration of algae and the tropic state of lakes correspond well to phosphorus levels in water. Studies conducted in the Experimental Lakes Area in Ontario have shown a relationship between the addition of phosphorus and the rate of eutrophication. Humankind has increased the rate of phosphorus cycling on Earth by four times, mainly due to agricultural fertilizer production and application. Between 1950 and 1995, 600,000,000 tonnes of phosphorus were applied to Earth's surface, primarily on croplands (Carpenter et al. 1998). Control of point sources of phosphorus have resulted in rapid control of eutrophication, mainly due to policy changes.

Human activities can accelerate the rate at which nutrients enter ecosystems. Runoff from agriculture and development, pollution from septic systems and sewers, and other human-related activities increase the flux of both inorganic nutrients and organic substances into terrestrial, aquatic, and coastal marine ecosystems (including coral reefs). Elevated atmospheric compounds of nitrogen can increase soil nitrogen availability.

Chemical forms of nitrogen are most often of concern with regard to eutrophication because plants have high nitrogen requirements so that additions of nitrogen compounds stimulate plant growth (primary production). Nitrogen is not readily available in soil because N_2, a gaseous form of nitrogen, is very stable and unavailable directly to higher plants. Terrestrial ecosystems rely on microbial nitrogen fixation to convert N_2 into other physical forms (such as nitrates). However, there is a limit to how much nitrogen can be utilized. Ecosystems receiving more nitrogen than the plants require are called nitrogen-saturated. Saturated terrestrial ecosystems contribute both inorganic and organic nitrogen to freshwater, coastal, and marine eutrophication, where nitrogen is also typically a limiting nutrient. However, in marine environments, phosphorus may be limiting because it is leached from the soil at a much slower rate than nitrogen, which are highly insoluble.

Ecological Effects

Adverse effects of eutrophication on lakes, reservoirs, rivers and coastal marine waters (from Carpenter et al., 1998; modified from Smith 1998). Many ecological effects can arise from stimulating primary production, but there are three particularly troubling ecological impacts: decreased biodiversity, changes in species composition and dominance, and toxicity effects.

Decreased Biodiversity

When an ecosystem experiences an increase in nutrients, primary producers reap the benefits first. In aquatic ecosystems, species such as algae experience a population increase (called an algal bloom). Algal blooms limit the sunlight available to bottom-dwelling organisms and cause wide swings in the amount of dissolved oxygen in the water.

Oxygen is required by all respiring plants and animals and it is replenished in daylight by photosynthesizing plants and algae. Under eutrophic conditions, dissolved oxygen greatly increases during the day, but is greatly reduced after dark by the respiring algae and by microorganisms that feed on the increasing mass of dead algae. When dissolved oxygen levels decline to hypoxic levels, fish and other marine animals suffocate. As a result, creatures such as fish, shrimp, and especially immobile bottom dwellers die off. In extreme cases, anaerobic conditions ensue, promoting growth of bacteria such as Clostridium botulinum that produces toxins deadly to birds and mammals. Zones where this occurs are known as dead zones.

New Species Invasion

Eutrophication may cause competitive release by making abundant a normally limiting nutrient. This process causes shifts in the species composition of ecosystems. For instance, an increase in nitrogen might allow new, competitive species to invade and outcompete original inhabitant species. This has been shown to occur in New-England salt marshes.

Toxicity

Some algal blooms, otherwise called "nuisance algae" or "harmful algal blooms," are toxic to plants and animals. Toxic compounds they produce can make their way up the food chain, resulting in animal mortality. Freshwater algal blooms can pose a threat to livestock. When the algae die or are eaten, neuro-and hepatotoxins are released which can kill animals and may pose a threat to humans.

An example of algal toxins working their way into humans is the case of shellfish poisoning. Biotoxins created during algal blooms are taken up by shellfish (mussels, oysters), leading to these human foods acquiring the toxicity and poisoning humans. Examples include paralytic, neurotoxic, and diarrhoetic shellfish poisoning. Other marine animals can be vectors for such toxins, as in the case of ciguatera, where it is typically a predator fish that accumulates the toxin and then poisons humans.

Nitrogen can also cause toxic effects directly. When this nutrient is leached into groundwater, drinking water can be affected because concentrations of nitrogen are not filtered out. Nitrate (NO_3) has been shown to be toxic to human babies. This is because bacteria can live in their digestive tract that convert nitrate to nitrite (NO_2). Nitrite reacts with hemoglobin to form methemoglobin, a form that does not carry oxygen. The baby essentially suffocates as its body receives insufficient oxygen.

Sources of High Nutrient Runoff

Characteristics of point and nonpoint sources of chemical inputs (from Carpenter et al, 1998; modified from Novonty and Olem 1994)

In order to gauge how to best prevent eutrophication from occurring, specific sources that contribute to nutrient loading must be identified. There are two common sources of nutrients and organic matter: point and nonpoint sources.

Point Sources

Point sources are directly attributable to one influence. In point sources the nutrient waste travels directly from source to water. For example, factories that have waste discharge pipes directly leading into a water body would be classified as a point source. Point sources are relatively easy to regulate.

Nonpoint Sources

Nonpoint source pollution (also known as 'diffuse' or 'runoff' pollution) is that which comes from ill-defined and diffuse sources. Nonpoint sources are difficult to regulate and usually vary spatially and temporally (with season, precipitation, and other irregular events).

It has been shown that nitrogen transport is correlated with various indices of human activity in watersheds, including the amount of development. Agriculture and development are activities that contribute most to nutrient loading.

There are three reasons that nonpoint sources are especially troublesome:

Soil Retention

Nutrients from human activities tend to accumulate in soils and remain there for years. It has been shown that the amount of phosphorus lost to surface waters increases linearly with the amount of phosphorus in the soil. Thus much of the nutrient loading in soil eventually makes its way to water. Nitrogen, similarly, has a turnover time of decades or more.

Runoff to Surface Water and Leaching to Groundwater

Nutrients from human activities tend to travel from land to either surface or ground water. Nitrogen in particular is removed through storm drains, sewage pipes, and other forms of surface runoff.

Nutrient losses in runoff and leachate are often associated with agriculture. Modern agriculture often involves the application of nutrients onto fields in order to maximise production. However, farmers frequently apply more nutrients than are taken up by crops or pastures. Regulations aimed at minimising nutrient exports from agriculture are typically far less stringent than those placed on sewage treatment plants (Carpenter et al., 1998) and other point source polluters.

Atmospheric Deposition

Nitrogen is released into the air because of ammonia volatilization and nitrous oxide production. The combustion of fossil fuels is a large human-initiated contributor to atmospheric nitrogen pollution. Atmospheric deposition (e.g., in the form of acid rain) can also effect nutrient concentration in water, especially in highly industrialized regions.

Other Causes

Any factor that causes increased nutrient concentrations can potentially lead to eutrophication. In modeling eutrophication, the rate of water renewal plays a critical role; stagnant water is allowed to collect more nutrients than bodies with replenished water supplies. It has also been shown that the drying of wetlands causes an increase in nutrient concentration and subsequent eutrophication booms.

Prevention and Reversal

Eutrophication poses a problem not only to ecosystems, but to humans as well. Reducing eutrophication should be a key concern when considering future policy, and a sustainable solution for everyone, including farmers and ranchers, seems feasible.

While eutrophication does pose problems, humans should be aware that natural runoff (which causes algal blooms in the wild) is common in ecosystems and should thus not reverse nutrient concentrations beyond normal levels.

Effectiveness

Cleanup measures have been mostly, but not completely, successful. Finnish phosphorus removal measures started in the mid-1970s and have targeted rivers and lakes polluted by industrial and municipal discharges. These efforts have had a 90% removal efficiency. Still, some targeted point sources did not show a decrease in runoff despite reduction efforts.

Minimizing Nonpoint Pollution: Future Work

Nonpoint pollution is the most difficult source of nutrients to manage. The literature suggests, though, that when these sources are controlled, eutrophication decreases. The following steps are recommended to minimize the amount of pollution that can enter aquatic ecosystems from ambiguous sources.

Riparian Buffer Zones

Studies show that intercepting non-point pollution between the source and the water is a successful mean of prevention (Carpenter et al., 1998). Riparian buffer zones, an interface between a flowing body of water and land, have been created near waterways in an attempt to filter pollutants; sediment and nutrients are deposited here instead of in water. Creating buffer zones near farms and roads is another possible way to prevent nutrients from traveling too far. Still, studies have shown that the effects of atmospheric nitrogen pollution can reach far past the buffer zone. This suggests that the most effective means of prevention is from the primary source.

Prevention Policy

Laws regulating the discharge and treatment of sewage have led to dramatic nutrient reductions to surrounding ecosystems, but it is generally agreed that a policy regulating agricultural use of fertilizer and animal waste must be imposed. In Japan the amount of nitrogen produced by livestock is adequate to serve the fertilizer needs for the agriculture industry. Thus, it is not unreasonable to command livestock owners to clean up animal waste-which when left stagnant will leach into ground water.

Nitrogen Testing and Modeling

Soil Nitrogen Testing (N-Testing) is a technique that helps farmers optimize the amount of fertilizer applied to crops. By testing fields with this method, farmers saw a decrease in fertilizer application costs, a decrease in nitrogen lost to surrounding sources, or both. By testing the soil and modeling the bare minimum amount of fertilizer needed, farmers reap economic benefits while the environment remains clean.

Natural State of Algal Blooms

Although the intensity, frequency and extent of algal blooms has tended to increase in response to human activity and human-induced eutrophication, algal blooms are a naturally-occurring phenomenon. The rise and fall of algae populations, as with the population of other living things, is a feature of a healthy ecosystem. Rectification actions aimed at abating eutrophication and algal blooms are usually desirable, but the focus of intervention should not necessarily be aimed at eliminating blooms, but towards creating a sustainable balance that maintains or improves ecosystem health.

HYPOXIA (ENVIRONMENTAL)

Hypoxia or oxygen depletion is a phenomenon that occurs in aquatic environments as dissolved oxygen (DO; molecular oxygen dissolved in the water) becomes reduced in concentration to a point detrimental to aquatic organisms living in the system. Dissolved oxygen is typically expressed as a percentage of the oxygen that would dissolve in the water at the prevailing temperature and salinity (both of which affect the solubility of oxygen in water; see oxygen saturation and underwater). An aquatic system lacking dissolved oxygen (0% saturation) is termed anaerobic, reducing, or anoxic; a system with low DO concentration-in the range between 1 and 30% DO saturation-is called hypoxic. Most fish cannot live below 30% DO saturation. A "healthy" aquatic environment should seldom experience DO less than 80%.

Causes of Hypoxia

Oxygen depletion could be the result of a number of factors including natural ones, but is of most concern as a consequence of pollution and as an outcome of a process known as eutrophication. Where plant nutrients enter a river, lake, or ocean, phytoplankton blooms are encouraged. While phytoplankton, through photosynthesis, will raise DO saturation during daylight hours, the dense population of a bloom reduces DO saturation during the night. When phytoplankton cells die, they sink towards the bottom and are decomposed by bacteria, a process that further reduces DO in the water column. If oxygen depletion progresses to hypoxia, fish kills can occur and invertebrates like worms and clams on the bottom may be killed as well.

Natural occurrences of hypoxia have been observed. Water flowing from a river into the sea is less dense than salt water. When this water does not mix with the underlying saline water, the oxygen concentration in the bottom layer may become low enough for hypoxia to occur.

Hypoxia is particularly problematic in shallow waters of semi-enclosed bodies of water like the Waddenzee or the Gulf of Mexico where land runoff is substantial. In these areas, a so-called "dead zone" can be created.

Solutions

To combat hypoxia, it is essential to reduce the amount of land-derived nutrients reaching rivers in runoff. Defensively this can be done by improving sewage treatment and by reducing the amount of fertilizers leaching into the rivers. Offensively this can be done by restoring natural environments along a river; marshes are particularly effective in reducing the amount of phosphorus and nitrogen (nutrients) in water.

In a very short time the oxygen saturation can drop to zero when offshore blowing winds drive surface water out and anoxic depthwater rises up. At the same time a decline in temperature and a rise in salinity is observed (from the longterm ecological observatory in the seas at Kiel Fjord, Germany). New approaches of longterm monitoring of oxygen regime in the ocean observe online the behavior of fish and zooplankton, which changes drastically under reduced oxygen saturations (ecoSCOPE) and already at very low levels of water pollution.

Bog Chemistry

In certain northern European sphagnum acidic bogs, a condition of hypoxia arises that prevents tissue decay by impeding micro-organisms in the soil and groundwater. Remarkable preservation of human mummies has occurred in some cases such as the discovery of Haraldskær Woman and Tollund Man in Jutland, Denmark.

MARINE POLLUTION

Marine pollution is the harmful entry into the ocean of chemicals or particles. A big problem is that many toxins adhere to tiny particles which are taken up within a few days by plankton and benthos animals, most of which are filter feeders, concentrating upward within ocean foodchains. Because most animal feeds contain high fish meal and fish oil content, toxins can be found a few weeks later in commonly consumed food items derived from livestock and animal husbandry such as meat, eggs, milk, butter and margarine.

One common path of entry by contaminants to the sea are rivers. The Hudson in New York State and the Raritan in New Jersey, which empty at the northern and southern ends of Staten Island, are a source of mercury contamination of zooplankton (copepods) in the open ocean. The highest concentration in the filter-feeding copepods is not at the mouths of these rivers but 70 miles south, nearer Atlantic City, because water flows close to the coast. It takes a few days before toxins are taken up by the plankton.

Many particles combine chemically in a manner highly depletive of oxygen, causing estuaries to become anoxic.

OCEAN ACIDIFICATION

Ocean acidification is the name given to the ongoing decrease in the pH of the Earth's oceans, caused by their uptake of anthropogenic carbon dioxide from the atmosphere. Between 1751 and 2004 surface ocean pH is estimated to have decreased from approximately 8.25 to 8.14 (Jacobson, 2005).

Carbon Cycle

In the natural carbon cycle, the atmospheric concentration of carbon dioxide (CO_2) represents a balance of fluxes between the oceans, terrestrial biosphere and the atmosphere. Human activities such as land-use changes, the combustion of fossil fuels, and the production of cement have led to a new flux of CO_2 into the atmosphere. Some of this has remained in the atmosphere (where it is responsible for the rise in atmospheric concentrations), some is believed to have been taken up by terrestrial plants, and some has been absorbed by the oceans. When CO_2 dissolves, it reacts with water to form a balance of ionic and non-ionic chemical species: dissolved free carbon dioxide (CO_2 (aq)), carbonic acid (H_2CO_3), bicarbonate (HCO_3) and carbonate (CO_3^{2-}). The ratio of these species depends on factors such as seawater temperature and alkalinity.

Acidification

Dissolving CO_2 in seawater also increases the hydrogen ion (H^+) concentration in the ocean, and thus decreases ocean pH. The use of the term "ocean acidification" to describe this process was introduced in Caldeira and Wickett (2003). Since the industrial revolution began, ocean pH has dropped by approximately 0.1 units (on the logarithmic scale of pH), and it is estimated that it will drop by a further 0.3-0.4 units by 2100 as the ocean absorbs more anthropogenic CO_2 (Caldeira and Wickett, 2003; Orr et al., 2005). Note that, although the ocean is acidifying, its pH is still greater than 7 (that of neutral water), so the ocean could also be described as becoming less alkaline.

Although the natural absorption of CO_2 by the world's oceans helps mitigate the climatic effects of anthropogenic emissions of CO_2, it is believed that the resulting decrease in pH will have negative consequences for oceanic calcifying organisms. These use the calcite or aragonite polymorphs of calcium carbonate to construct cell coverings or skeletons. Calcifiers span the food chain from autotrophs to heterotrophs and include organisms such as coccolithophores, corals, foraminifera, echinoderms, crustaceans, and some mollusks, especially pteropods. Aside from calcification (and specifically calcifiers), organisms may suffer other adverse effects, either directly as reproductive or physiological effects (e.g. CO_2-induced acidification of body fluids, known as hypercapnia), or indirectly through negative impacts on food resources. However, as yet there is not a full understanding of these processes in marine organisms or ecosystems.

Under normal conditions, calcite and aragonite are stable in surface waters since the carbonate ion is at supersaturating concentrations. However, as ocean pH falls, so does the concentration of this ion, and when carbonate becomes undersaturated, structures made of calcium carbonate are vulnerable to dissolution. Research has already found that corals (Gattuso et al., 1998), coccolithophore algae (Riebesell et al., 2000) and pteropods (Orr et al., 2005) experience reduced calcification or enhanced dissolution when exposed to elevated CO_2. The Royal Society of London published a comprehensive overview of ocean acidification, and its potential consequences, in June 2005 (Raven, et al., 2005).

While the full ecological consequences of these changes in calcification are still uncertain, it appears likely that calcifying species will be adversely affected. There is also some evidence that the effect of acidification on coccolithophores (among the most abundant phytoplankton in the ocean) in particular may eventually exacerbate climate change, by reducing the earth's albedo as well as oceanic cloud cover (Ruttiman, 2006). Present evidence suggests that dramatic changes in the biogeochemistry of the marine environment over the next 100-200 years can be avoided only with early and deep reductions in carbon dioxide emissions.

OIL SPILL

An oil spill (sometimes called an oil slick) is the unintentional release of liquid petroleum hydrocarbon into the environment as a result of human activity. The term often refers to marine oil spills, where oil is released into the ocean or coastal waters. Oil can refer to many different materials, including crude oil, refined petroleum products (such as gasoline or diesel fuel) or by-products, ships' bunkers, oily refuse or oil mixed in waste.

Oil is also released into the environment from natural geologic seeps on the seafloor, as along the California coastline. Most man-made oil pollution comes from land-based activity, but public attention and subsequent regulation has tended to focus most sharply on seagoing oil tankers.

The fate, behavior and environmental effects of spilled oil can vary, depending upon the type and amount of material spilled. In general, lighter refined petroleum products such as diesel and gasoline are more likely to mix in the water column and are more toxic to marine life, but tend to evaporate relatively quickly and do not persist long in the environment. Heavier crude or fuel oil, while of less immediate toxicity, can remain on the water surface or stranded on the shoreline for much longer. Oil from the Exxon Valdez and Gulf War oil spills, while weathering over time, has persisted along the shoreline for years after the spill. By contrast, the Braer spill off the Shetland Islands in 1993 and the Sea Empress spill off Milford Haven in 1996 left almost no long-term environmental damage, despite both being twice as large as Exxon Valdez.

With vast numbers of people living and depending on coastal areas for fishing and tourism throughout the world, the consequences of oil spills can be serious. Such possibilities have caused outcries for oil companies, ship owners, and shipbuilders to share the responsibility of preventing such a disaster. As oil is lighter than water, and does not decompose, it will remain on the surface for a very long time. As it is also flamable, oil spills can often fuel ocean fires.

Ships today are better equipped and built than ever before, with all new-build large tankers having double hulls. Ship-source pollution, which averaged over 400,000 tonnes a year at its peak in the late 1970s, averaged 27,000 tonnes per year from 2000-05, a fall of 93%.

Another way to mitigate the effects of an oil spill is through being able to contain and effectively treat spilled oil. A new vital step being developed in preventing oil damage is through installations of systems made for Fast Oil Recovery (FOR) of oil from wrecked ships.

Environmental Effects

Studies of the Exxon Valdez oil spill have shown that the environmental damage caused by oil spills can be greater than was previously thought. It is now thought that the impacts to marine life can be evident at less than one part per billion petroleum hydrocarbons. The lighter fractions of petroleum, such as benzene and toluene, are more toxic, but are more volatile and evaporate quickly. Heavier components of crude oil, such as polynuclear aromatic hydrocarbons (PAHs) appear to cause the most damage; while they are less toxic, they persist in the environment much longer than volatile components.

TYPES OF POLLUTION

Anything that dissolves in water, or can be suspended in water, or floats on water can be considered a pollutant. However, there are eight primary types of pollution that account for most of the problems that humans produce. These are water pollution by:

Sediment

Sediment is the primary water pollutant in the United States. Sediment is undissolved solid material (sand, silt and clay) that has been eroded from the land. Such materials are washed away from all land areas during rain storms, floods and snow melts, but the volume of sediment becomes much greater if some of the land's natural protections have been removed. This is often the case on construction sites, in agricultural areas, and after forest fires. Sediment gives water a cloudy appearance, which is called "turbidity." Turbid, or muddy, water does not allow sunlight to penetrate; it can hamper the ability of aquatic animals to get oxygen out of the water; and it can bury bottom-dwelling organisms as it settles out of the water. Sediment has the additional problem-causing characteristic that many pesticides, nitrates and pathogenic organisms attach themselves to particles of silt and clay.

Toxic Matter

A great variety of poisonous compounds can get into water either by accident or intent. For example, humans have known for years that lead is a poison that affects the nervous system, but we realized only recently that the lead in plumbing and the lead in automobile gasoline can enter our water supply and harm us. There discovery gives us a good reason to believe that the Romans may have been harmed by lead in their water and wine over 2000 years ago. Another example of toxic material in water occurred in the 1950s when industries around Minamata Bay in Japan discharged so much mercury into the bay that people were poisoned by mercury that had found its way into the local seafood. More recently, people have become concerned about the presence of agricultural nitrates in water. The human body converts nitrates into nitrites, and these can interfere with the blood's ability to transport oxygen. In these and many other cases, material with one purpose (plumbing, industrial production or fertilizer) ended up in the wrong place as a serious water pollutant.

Pathogens

Pathogens are agents that cause disease. They may be bacteria, viruses or parasites, and they can get into water from any infected organisms. Small numbers of these agents can be found in almost all bodies of water, but the numbers greatly increase when any sort of animal waste enters the water. In parts of the world where untreated sewage is allowed to enter the water supply, contaminated water is a major health hazard.

Salt

Most land plants and animals can be harmed by salt (sodium chloride and calcium chloride), and therefore any salt in the supply of fresh water can be a problem. Under certain conditions, salt water from the ocean can intrude into groundwater, and storms can blow ocean water into surface lakes and streams. Irrigation projects can add salt to streams, and salt water is a common by-product of the oil industry. (Crude oil often comes to the surface mixed with salt water). In the northern parts of North America, salt is spread on roads and bridges to stimulate the melting of ice. Salt from storage areas as well as from the highways eventually finds its way into local streams and groundwater.

Acid

Acid mine drainage (acids given off by the weathering of ores) is a significant source of water pollution in mining areas, and acid precipitation is becoming an important problem in industrialized areas. In addition, acid can enter the water from a wide variety of industries (such as steel-making and fruit-packing) that use acids as cleaners. In all cases, the addition of acid to water lowers its pH and often kills the plants and animals that would naturally inhabit it. Acidic water also dissolves metals more easily, and thus the water may become additionally polluted by metals such as lead, copper, or mercury.

Agricultural Chemicals

Most agricultural chemicals are water-soluble nitrates and phosphates that are applied to fields, lawns and gardens to stimulate the growth of crops, grass and flowers. The chemicals enter streams and lakes during the run-off of rain water. Once in a body of water, these chemicals continue to promote the growth of plants, but an overgrowth of plant life can seriously damage a lake or stream. The resulting plant detritus is food for micro-organisms, and as the population of such organisms grows, the supply of oxygen in the water is depleted. Declining oxygen leads to a drop in the population of fish. In extreme cases, the plants may produce so much detritus that all oxygen is used up and all animal life disappears from the stream. Water in which this problem is occurring, is said to have a high "Biochemical Oxygen Demand", or "BOD". This means that the water is capable of supporting a large population of bacteria that will have a high demand for oxygen.

Organic Matter

Dumping untreated sewage or other organic material into a body of water has much the same effect as pollution by fertilizers. In both cases the water is supplied with an abundance of food for micro-organisms, and the rapidly expanding population of microscopic animals soon depletes the supply of oxygen making the water uninhabitable by fish and other larger organisms.

Heat

People use a large amount of water for cooling. This water is usually returned to its source only about 10 degrees warmer than its natural temperature, but even this amount of warming can affect the water's ability to support life. The problem is known as "thermal pollution." Cold water holds more oxygen than warm water. For example: at 32° F, fresh water can hold 14 ppm (parts per million) of oxygen. If that same water is warmed to 70° F, it can only hold 9 ppm oxygen, and at 100° F, its oxygen content decreases to 6 ppm. As the water's oxygen supply decreases, it loses its ability to support animals such as fish. The problem is made worse by the fact that warmer water encourages the growth of bacteria that may attack the fish. Thermal pollution is particularly severe during summer months when the water's natural temperature is high. During very hot weather, a body of water may hold only 50% as much oxygen as it would during the winter. Artificially warming the water at that point could easily begin to kill off the fish.

Point Sources

Pollutants that are discharged at a known location through conduits such as sewers, ditches or canals, are said to be coming from a "point source." Such sources are easy to locate, and they are also easy to regulate and control. It is possible to identify the polluting material and determine its volume. It is also easy to determine the extent to which regulations are being followed.

Non-Point Sources

"Non-point sources" are less distinct sites such as lawns, forests, construction sites, parking lots, or farms. Overall, agriculture is the leading source of non-point pollution in the United States. Effluent from agricultural sources produces about 65% of the total mass of pollutants entering the nation's streams and about 55% of those entering lakes. Non-point sources are much more difficult to control since it is often impossible to identify the nature and the volume of the material entering the water supply.

Groundwater

All groundwater was surface water at one time, and therefore, anything that is a potential threat to surface water is also a threat to groundwater. Groundwater can become polluted, and once it does, it may take thousands of years for the contaminants to be flushed from the aquifer. Pollutants can be washed out of rivers in weeks or months, and lakes can clear themselves out in decades; but reservoirs of groundwater may remain contaminated for millennia. Since most groundwater is invisible, it may even take years just to detect a problem. Farming and gardening use pesticides, herbicides and chemical fertilizers that may enter the ground. Leaking underground storage tanks at homes and service stations can pollute acres of groundwater with various petroleum products. Groundwater can also be contaminated by organic waste from outhouses, septic tanks, and leaking sewage lines. All such pollution must be located, and some means devised to remove it from the ground.

Solutions

At first look, solutions to water pollution seem remarkably simple. Either prevent the problem from happening, or clean up the pollution that has occurred. Problems are handled in both of these ways, but solutions become complicated when they are put into practice. Prevention and clean-up can each be time-consuming and expensive. The source of the pollution must be located and identified; an alternative method of dealing with the waste must be determined; and often, time must pass while the pollutants are "flushed" out of the ground or the effected body of water.

Prevention

Legislation, such as the Safe Drinking Water Act in the U.S., is aimed at the prevention of pollution. The goal is to eliminate practices that allow pollutants to enter the water supply in the first place. Some pollutants are toxic or disease-causing materials that should not be released under any conditions. Other pollutants are only harmful in quantities that are too large to be handled by natural systems. The appropriate legislation may be as simple as regulations about dumping used motor oil onto the ground or into sewage systems, or it may be as complex as the regulations controlling the effluent from industries or municipal sewage treatment facilities. In each case, the intent is to remove toxic material from the environment, and to limit the rate that harmful materials are added to water supplies so that natural processes can absorb or destroy them.

Clean Up

A salt marsh may not look like a pollution control device, but it contains many physical and biological processes that work to remove polluting material from water. These natural processes allow water to be considered a "renewable resource" in that the water may serve one purpose (such as washing nutrient-rich mud into the marsh), and it may then be cleaned and re-used in another capacity (such as drinking water for a bird). Humans try to duplicate some of these natural processes in order to clean up water that we have polluted.

Natural Processes

There are four significant natural processes that act to clean water and allow it to be reused. They are:

- *Evaporation:* When water turns from a liquid into a vapor, almost all of the material that was dissolved or suspended in the water is left behind. As a result, salt water evaporates to create a vapor of fresh water, and muddy water evaporates to create a vapor that does not contain any mud.
- *Sedimentation:* Various types of sediment are the most common agents of pollution in water. Sediment is suspended in water by turbulence, and slowly settles out of water that is allowed to sit still. The larger particles of sediment drop to the bottom first, but if a sample of water remains still for a long period of time, even the smallest particles of clay will settle to the bottom.
- *Adhesion:* Some natural materials, such as clay, adhere readily to other materials. The process is known as "adhesion," (as in adhesive tape) and results in the fact that some of the pollutants in water stick to particles of clay and fall to the bottom as the clay settles. Almost like a natural filter, this process of adhesion can remove dissolved as well as suspended pollutants from water. This process is primarily responsible for the generally correct belief that a stream will "clean itself" if it is allowed to flow far enough.
- *Biologic Activity:* Almost all of the organic waste from one organism is food for another. Dead leaves and branches are consumed by micro-organisms; the waste from the micro-organisms is then used by plants to form

leaves and branches. If not disturbed, this natural cycle can operate continuously-consuming potentially polluting wastes and reusing them in the natural system. Many of these same natural processes are used by humans to clean-up our own water pollution problems.

Human Processes

Humans take advantage of natural cleaning processes to deal with many water pollution problems. In two of our most significant processes, we use "sedimentation and adhesion" to treat drinking water, and we take advantage of "sedimentation and biologic activity" in the treatment of sewage.

Municipal Water Treatment

If a city's supply of water is drawn from lakes or rivers, the primary goal of the municipal treatment facility is to clean the water of sediment and pathogens. Sediment is removed by stimulating the small particles of clay that are suspended in the water to stick together in larger groups.

Once the groups are large enough, they fall to the bottom and are removed. This process is known as "flocculation." Flocculation is a natural process, but it is made to operate more quickly in most water treatment plants by adding "alum" (aluminum sulfate) to the water. After the flocculation process, the water passes though fine "sand filters," which often contain activated charcoal as well.

This sand, or sand/charcoal combination, filters out the smallest particles and also removes small amounts of organic matter that might give the water a strange taste or smell. Pathogens are removed in two ways. First of all, chlorine is added to the water as it enters the plant, and as it leaves. The chlorine added before treatment will kill bacteria and fungi before they enter the plant, and the chlorine added at the end of the process will protect the water as it is piped to consumers. Second of all, many pathogens and chemical impurities are removed by the natural process of adhesion to clay particles that are settling to the bottom of the treatment tanks. Once the water has been cleaned, fluoride and several other chemicals are added before it is pumped out of the plant.

Municipal Waste Water Treatment

The mission of a municipal waste water treatment plant is to clean the waste water sufficiently so that it can be returned to a river or lake without altering the natural conditions in that body of water. To do this, the plant must first remove debris that would spoil the appearance of the water. In addition, the plant must remove pathogens from the water, and it must remove organic matter that would promote the growth of plants and eventually deplete the water's supply of oxygen. Early waste water treatment facilities were "Primary" systems that used only physical means to clean the water. Screens were used to remove large debris, and "clarifers" (settling tanks) were employed to allow smaller suspended or floating solids to separate from the water. The cleaner water was then chlorinated and released. Today, most major U.S. cities operate "Secondary" systems that use both physical and biological processes to clean the water. In addition to the primary treatment described above, these plants also have "aeration basins" in which oxygen is added to the waste water to encourage bacteria to consume suspended and dissolved organic matter. Good secondary systems can significantly reduce the waste water's BOD. In places where waste water contains many chemical impurities, or where it must be cleaned to a very high standard, various types of "Tertiary," or chemical, treatment are used. This type of treatment is often expensive and its use is limited.

Soil Pollution

Soil pollution is any chemical or biological alteration of soil that is harmful to living organisms. Virtually all soil pollution is the direct result of sources of pollution on the ground surface.

These percolation of these surface pollutants through the soil cause soil degradation, but also acts as a natural cleansing process for ground water. However, solutions, through regulations, are feasible.

Sources of Soil Pollution

Pollution on the ground surface is the cause of most soil pollution. This surface pollution comes from many sources, but these can be grouped into three categories.

Waste Disposal Practices

The disposal of three sources and types of wastes are one of the primary sources of pollution on or near the ground surface. The disposal of these wastes are discussed in these sections.

Hazardous Waste

When we think of hazardous waste, the image of industrial by-products stored in rusty barrels generally comes to mind. However, hazardous materials are also found in cleaning closets, under the sinks, and on the garage shelf in our homes. Agricultural by-products are considered hazardous materials. These substances all pose special disposal concerns. Hazardous waste includes any toxic material whose disposal can endanger people or the environment. Although there are some naturally occurring substances that pose health risks if not properly handled, by far the greatest threat comes from three sources: synthetic chemicals, pesticides and toxic household products. We come in contact with these materials in a great variety of ways.

Sources

Synthetic chemicals are used to create the majority of the products in industrial societies. During the manufacture of consumer products, the possibility exists for toxic materials to enter the environment via smokestack emissions, from dumping into surface waters, from leachate lost from landfills and finally as leaked contaminants from deep disposal wells. Most modern agriculture uses a large number of pesticides and herbicides. These synthetic products get into the soil and then into the hydrosphere by runoff from the fields. Toxic household products include cleansers, detergents, insecticides, used motor oil, and batteries of all sorts. Many people are unaware of the dangers these products pose, and unless communities have special disposal programs for the materials, these products end up being poured down drains and dumped into municipal landfills. There they contaminate the water, because waste treatment plants are not equipped to remove contaminants, or they leak out as leachate from landfills which are not designed to contain such chemicals. In both cases the wastes become significant threats to the environment.

Threats

Plants and animals have no natural resistance to many synthetic products, and thus their presence in soil and water upsets the balance of ecosystems. Pesticides, designed to eliminate one insect, may also kill that insect's natural predators thus increasing a problem instead of decreasing it. The insecticide may also kill or threaten a large number of other animals such as fish or birds that were never intended to be targets. Contaminants in water can produce lesions on fish or kill fish, and these are often only the very visible signs of more extensive disruption of the ecosystem.

The effects of hazardous waste contamination on humans includes a wide variety of symptoms which depend on the type of pollutant and the dose received. In extreme instances the toxins can maim or kill a person instantly, in other cases they can produce brain damage, hearing loss, behavior problems, cancer, genetic mutations, and birth defects and miscarriages. In an attempt to avoid some of these problems, people can deal with hazardous waste in two basic ways. One method is to find safe ways to dispose of such waste. Another approach is to find ways to decrease the amount of waste produced.

Disposal

Secure Landfills: One method used for the disposal of hazardous waste is to place it in secure landfills. Although some might argue that there is no such thing as a secure landfill the idea behind them is to prevent any leachate from escaping from the disposal site into local aquifers. This is achieved by using geologic barriers such as the type of rock and soil in the disposal area, and by lining the site with an impermeable plastic sheet. Materials are placed in special compartments in the landfill to prevent any undesirable mixing of the different substances. After it is filled, the site would be capped with a layer of impermeable clay to cut down on the amount of water that will pass through. Test wells are drilled into the site to detect any leaking. The main disadvantage to this method of disposal is its failure rate. It is very difficult to isolate the material from the groundwater, and sites which were thought to be absolutely leak-proof have leaked.

Disposal Wells: Another method to dispose of liquid hazardous waste is to pump it deep into the ground. The objective is to isolate the waste from the surface aquifers by placing it below them. Only some areas meet the geologic requirements necessary for this type of disposal. There must be a geologic layer (usually sandstone) at depth to receive the waste, which is separated from the surface aquifers by impermeable layers (usually clay or shale). The waste is injected to the layer through special wells lined with cement. The risks associated with this method of disposal involve leaking at the well entrance, or undetected fractures in the rocks, which would allow contamination of the surface aquifer.

Incineration: A final method to dispose of hazardous waste is to incinerate the material. As long as it is carried out at a sufficiently high temperature, incineration has the advantage of destroying the hazardous nature of the material. Another advantage of this method is that it reduces the volume of the waste material. Suitable for solid and liquid waste, incineration breaks up organic compounds into harmless elements which combine with oxygen and are released into the atmosphere. In some cases, toxic ashes, acids, or metals are the only materials left over to be handled after incineration. The associated risks of incineration are inadvertent leaking of toxic ash or gasses during the process itself. Also, CO2 is emitted as an end product of the process which, though not toxic, is a greenhouse gas.

Minimization

Part of the solution to the problem of hazardous waste has to be to reduce its production. One way to reduce production is to change manufacturing to utilize "cleaner" processes. Another method is participation in waste exchange programs, where the waste from some companies is used as a resource for other companies. Many wastes can be re-processed and reused. One of the (outwardly) simplest ways to minimize waste production is to reduce demand for consumer goods. If the goods are not produced, neither will the wastes be produced.

Nuclear Waste

Today there are many uses for nuclear technology which range from simple detectors to the construction of complex weapons. All applications of nuclear technology produce materials which are no longer useful, but are radioactive. These waste materials must be disposed of in a manner which will minimize human exposure to radiation until the material is no longer dangerous.

Nuclear Energy

Nuclear energy is derived from atoms when their nuclei are split to form less complex atoms, or combined to form larger ones. This change can occur spontaneously when naturally unstable atoms come apart and emit particles and energy, or it can be made to occur on command in nuclear reactors. Humans use nuclear reactions to generate electric power and for a variety of other purposes. In most cases, the atoms created by nuclear change are radioactive, meaning that they are unstable material which will continue to emit particles and energy until it has decayed to a stable state. The energy and particles which are given off are hazardous to most living things.

The time it takes the atoms of a particular element to decay to a stable form is expressed in "half-lives". An element's half-life is the time it takes for half of the material to change to its stable form. An element can remain dangerously radioactive through many half-lives. Materials with short half-lives (measured in seconds) lose their radioactivity quickly, and those with very long half-lives (measured in millions of year) are emitting radiation very slowly. As a result, neither of these poses as much of a threat to humans as does material with an intermediate half-life which may remain dangerously radioactive to 100's or 1000's of years. These materials must be safely isolated from the environment for long periods of time.

Sources

The chief uses of nuclear technology include medical diagnosis and treatment, food preservation, electricity generation, weapons production, and the general research which supports those industries. In medicine, radioactive substances are used for diagnostic purposes and for the treatment of cancer. Food irradiation can kill bacteria, insects and fungi without making the food itself radioactive. Nuclear power plants produce electricity for residential and commercial use. The military designs and tests nuclear weapons, and uses nuclear fuel for ships and submarines. All these endeavors produce radioactive waste; some liquid, some solid, and some in the form of a gas. A certain percentage of this waste is highly radioactive and very dangerous, this is classified as high-level waste. Much of the rest of the waste is mildly radioactive and is considered low-level waste.

Low-Level Waste

Low-level waste includes materials such as contaminated clothing, water used for cleaning, filters and some medical waste. It also includes gas emissions from nuclear plants. These may contain only small amounts of radioactivity, but this type of waste accounts for a large percentage of nuclear waste. Much of this material is taken to commercial disposal sites.

Commercial Disposal Sites

For gases and liquids which contain low-levels of radiation, the main disposal objective is to dilute the material to low concentration levels so that it poses a minimal threat to the environment. The gasses are released into the atmosphere in small quantities. Liquid waste is diluted to a level which is deemed safe and then released.

High-level Nuclear Waste

High-level nuclear waste is material with high radioactivity and a long half-live. It consists of spent fuel rods, and other by-products from the operation of nuclear reactors. Some of the material is toxic, as well as being radioactive, so extreme caution is required in managing this waste. Most people believe that the only safe way to "dispose" of such material is to find a place where it can be permanently isolated from the environment.

Permanent Storage

Today most high-level waste is temporarily stored at its site of origin. The accompanying map shows the approximate locations of the waste. The objective is to place it in a location in which it is isolated from the hydrosphere, atmosphere, lithosphere and biosphere long enough for the radioactivity to diminish (tens of thousands of years). With the dynamic nature of the Earth this is a difficult task.

Spills

Soil contaminants are spilled onto the surface through many different activities. Most of these are the result of accidents involving the vehicles that are transporting waste material from the site at which it originated to the site at which it is to be disposed. Others involve accidents involving vehicles (automobiles, trucks and airplanes) not transporting wastes, but carrying materials, including fuel, that when spilled contaminate the soil. Still other spills are the direct action of humans pouring potentially toxic materials (solvents, paints, household cleaning agents, etc.) onto the ground surface rather than disposing these materials by more appropriate means.

AGRICULTURAL PRACTICES AND CHEMICALS

Agricultural practices, including the use of agricultural chemicals, are another primary source of pollution on or near the ground surface. The specific practices and their resultant pollution are each discussed in the sections on:

Percolation of Surface Pollutants through Unsaturated Soil

When any liquid pollutant is on or just below the ground surface for any period of time, one of three things could happen to it, if it is not cleaned up first. First, the pollutant might be washed away by precipitation, causing little or no harm to the ground on which it was found. This, however, is not a solution to the bigger pollution problem, because the pollutants will simply accumulate somewhere else. Second, the pollutant, if volatile, could evaporate, again causing little or no harm to the ground on which it was found. Again, however, this is not a solution to the bigger pollution problem, as it might become a source of air pollution. Third, the pollutant could infiltrate through the unsaturated soil, in much the same way as ground water. Percolation of surface pollutants through soil have two major effects, soil degradation, which is negative and cleansing of ground water, which is positive.

Soil Degradation

As liquid pollutants flow through unsaturated soil, some of the pollutants are removed by the soil particles. This is especially true if the soil contains much clay, because many pollutants will adhere to the clay particles. Other pollutants could enter into chemical reactions with soil particles, effectively removing them from the liquid, but storing them in the soil. A major cause of concern over soil pollution is that soil provides nutrients to growing plants, including food, and serves as a habitat for many animals. If soil is polluted with agents harmful to food plants, and if those pollutants enter into the plants, it renders the soil unfit for the purposes of growing food. If soil is polluted with agents harmful to the animals living in the soil, it renders the soil unfit for habitation. This is of particular concern because it could throw the whole ecosystem out of balance.

Natural Cleansing Process for Ground Water

One good effect comes from the passage of liquid pollutants through unsaturated soil, especially if the pollutants are being carried in water. Percolation of water from the surface through the unsaturated soil zone can have a cleansing effect on the water before it becomes ground water. As discussed in the section on soil degradation, soil particles have the ability to remove some contaminants from water.

Solutions

All of the above problems are very serious. However, all of the problems have solutions. Regulations dealing with the disposal of wastes, clean up of spills and agricultural practices are in effect, but can be strengthened and must be enforced. A more positive environment for the establishment and enforcement of more rigorous regulations can be fostered through public education.

The general public must become aware of the causes and especially the effects of pollution. Once the public becomes concerned about these issues, they will demand that their elected officials create and enforce legislation designed to improve the situation.

EXTINCTION

Extinction is the complete and permanent disappearance of a species from earth. People tend to look upon extinction as a tragedy or a disaster, because we have seen many species of plants and animals disappear over the last couple of centuries. We realize that humans are largely responsible for these recent extinctions because human activities have changed the surface of the earth in ways that make it difficult for many species to survive.

In the natural world, however, extinction is seen as the partner of evolution in the development of living things. Each process is a response to natural events that change the environment. When conditions change, a species can either evolve and adapt to suit the new conditions, or it can die off and become extinct. No matter how successful a species might be, its ultimate fate is to become extinct; just as any plant or animal must eventually die. Biologist estimate that well over 99% of all species that have ever existed on earth are now extinct. Most of these extinctions occurred long before people were present on earth. According to evolutionary biologist David Raup, species have a lifetime of about 4-10 million years. Since life has existed on earth for nearly 4 billion years, the majority of species have developed, flourished for a time, and become extinct long ago.

Throughout most of the earth's history, extinction has been a gradual process with perhaps 10 to 20 species disappearing each year. Occasionally, a rapid and wide-ranging change in conditions could bring about a mass extinction involving a significant percentage of the species existing at the time. A well known example is the mass extinction 65 million years ago that wiped out the dinosaurs and many other animals and plants. Such an event, of course, leaves the door open for the evolution of the remaining plants and animals to fill the habitats left open by the extinction.

Mass excitations in the past have been the result of natural events, but a mass extinction is occurring today that is primarily the result of human activities. There have been estimates of the current rate of extinction that vary from several hundred per year to several tens of thousands per year. An average of present estimates indicates that about 100 species are becoming extinct every day, or about 36,000 each year. Should people be concerned about this rate of loss? What, if anything can be done about it?

NATURAL EVENTS

Most natural extinctions are the result of changes in climate. Basically, an area's climate is defined by its range in temperature and the amount of moisture it receives. Changes in this climate can be gradual or rapid. Gradual changes that take place over millions of years are generally related to the motion of the earth in its orbit around the sun, or to the motion of the earth's crustal plates. Rapid changes may take place in a matter of years or hundreds of years, and are usually related to dramatic geologic events such as a volcanic eruption or the impact of a meteorite. However rapidly a climate may change, its alteration can change all of the major aspects of an environment, and therefore determine which species can survive and which and cannot. For example:

Growing Conditions-Climate has a direct control on the type and number of plants that can live in a particular area. Most plants require a specific range of temperature and moisture to survive. If they cannot adapt to a change, they will die out.

Supply of Food: All animals are ultimately dependent on plants for food. The animals may eat the plants directly, or eat other animals that eat plants. In either case, a change in climate can drastically change the type and amount of food available.

Predators: A change in climate might favor one species that is a predator for another with the result that the pray are driven from the area or killed off.

Disease: A species of plants or animals may be weakened by some of the changes listed above, or a pathogen may be strengthened so that the plants or animals are overtaken by disease.

Competition: A different climate may weaken certain organisms, or strengthen others, so that one my be able to compete more successfully for a particular niche in the environment.

HUMAN ACTIVITIES

Mass extinctions have occurred many times in the past, but the current extinctions are the first that can be attributed to the actions of a single species, humans. In a broad sense our effect is due to three factors:

(1) The growth in our population,

(2) Our policies that value our own well being over that of the environmental system that support us, and

(3) Our ever increasing appetite for natural resources.

More specifically, our effect on the rate of extinction can be attributed to the following:

Destruction and fragmentation of habitat-Since humans first began to farm, we have been disrupting natural lands in order to supply ourselves with food and other resources. Logging, road construction, housing, and mining all destroy or subdivide former habitats for wildlife. In the 48 contiguous United States 98% of the tall-grass prairies have been plowed, 90% of old-growth forests have been cut, and the total amount of forest cover has been reduced by 33%. It has been estimated that these changes have led to the extinction of at least 500 species. National Parks and other preserves serve as small protected habitats, but often the space is too limited to support a breeding population, or the distance between safe areas is too great to allow growing populations to colonize new area and return to breed.

- *Commercial Hunting and Poaching:* International trade in exotic plants and animals is a large and lucrative business. A live mountain gorilla is worth $150,000, a Bengal tiger skin sells for $100,000, and rhinoceros horn will bring $28,000 per kilogram. Over 600 species of plants and animals are severely threatened because of illegal trade. Unfortunately most penalties are small and enforcement is difficult. The U.S. Customs Service believes it catches less than 10% of the illegal wildlife trade coming into this country.

- *Overfishing:* Humans have become very adept at catching fish. As our boats, nets, and location devices have become more sophisticated; the fish have become more scarce. From 1950 to 1990 the worldwide harvest of fish increased from 20 million tons per year to 100 million tons. Despite a continuing increase in the size and power of the world's fishing fleets, the catch during the 1990 has leveled off. The UN Food and Agricultural Organization estimates that since 1993, 15 of the world's 17 major fishing areas are being fished at or beyond their estimated sustainable yields.
- *Control of Predators and Pests:* For years people have tried to eliminate plants and animals that compete with us for food in one way or another. In the U.S. 99% of the prairie dogs have been eliminated because their borrows injure horses and cattle. In Africa large numbers of elephant are killed because they trample crops. The Carolina parakeet was eliminated because it ate fruit, and wolves and coyotes have been killed because the attach livestock.
- *Pollution:* Pollution of air, water, and soil has been harmful to wild plants and animals just as it has been to humans. Some air pollution is believed to have a direct effect on climate which is the primary cause of extinction.
- Non-Indigenous Species

According to several studies, approximately 32% of the known species in the US are vulnerable to extinction as a result of human activities. The 10 states in which the treat is particularly severe include California, Hawaii, Texas, Florida, and most other southeastern states.

Solutions: There are basically two ways to protect wildlife. Most biologists believe that the most effective approach is to protect the ecosystems in which the plants and animals belong. This involves establishing and protecting natural wildlife areas in which the native species are balanced and preserved, while the non-native species are removed. The primary problem with this system is that only 6% of the world's land area is currently protected as wildlife sanctuary, and with a rapidly growing human population, is it not likely that other land can set aside for this purpose.

The more common method to protect wildlife is to identify and protect specific species that are at risk. A species is said to be "endangered" if so few individual exist that it may become extinct over all or some of its natural range. Examples of endangered animals are the condor in California, the black rhinoceros in Africa, and the giant panda in China. A species is classified as "threatened" if it is still abundant in its natural range, but its numbers are declining such that it may become endangered. Examples of animals in this category include the bald eagle, the grizzly bear and the alligator. The Convention on International Trade in Endangered Species was established in 1975 and lists over 600 species that cannot be commercially traded as specimens or products because they are threatened or endangered. In the United States, the Endangered Species Act of 1973 (amended in 1982 and 1988) is one the world's toughest environmental regulations. The act does not allow Americans to import any product made from an endangered or threatened species. In addition, the act authorizes the U.S. Fish and Wildlife Service and the National Marine Fisheries Service to identify threatened and endangered species that cannot be hunted, killed, collected or injured in the United States. To date about 960 species have been identified. Once a species is identified, the Endangered Species Act requires that a plan be developed to help the species recover. Due to lack of funding, only about 600 plans have been devised, and most of these have not yet been put into practice.

Throughout most of its existence, the Endangered Species Act has been under pressure from groups that believe it unfairly restricts their ability to do business. Many logging companies, developers, miners, and others argue that the Act is ineffective and an unnecessary expense since it has removed only seven species from the list and only 20 have recovered sufficiently to be reclassified from endangered to threatened. Senator Slade Gordon from Washington has suggested that all endangered species should be removed from the wild and bred in zoos so that endangered plants and animals can be protected without blocking economic development. Those who support the Act point out that recovery is a slow process, and 20 years is a very short time in which to expect results. Supporters also note that the total budget for the Endangered Species Program in the U.S. in 1996 was $57 million, about equal to the cost of 1.5 miles of urban interstate highway.

Concerns: If extinction is a natural process, and the competition between species for space and food is natural as well, why should we be concerned if humans are currently winning the battle against almost all other organisms and some of them are becoming extinct? The answer, according to many people, has three parts. The first is very selfish. The vast majority of our food, shelter, and medicine has been derived from wild plants. Many wild plants and animals recycle nutrients, build soil, and remove toxic materials from the air and water. So far we have examined very few of the thousands of known species. If we allow them to become extinct before we know how important they are to our survival, we may lose a great deal. The second is slightly less selfish. Humans enjoy wildlife and spend millions of dollars each year to visit natural areas, see beautiful plants, and watch interesting animals. Biologist Edward Wilson refers to our affinity for nature as "biophilia," and claims that it is an imprint in our genes that reflects our "roots in nature." The third part of the answer is that as intelligent and thinking organisms, people are obliged to look beyond their immediate needs to the rights of other organisms. Many people believe that all species have an equal right to compete for a place on earth, and we are obliged to make sure that the competition is fair.

SUSTAINABILITY

There are two very different ways of looking at the relationship between Planet Earth and its human inhabitants-the Frontier World-View and the Sustainable World-View. Each of us views this relationship from a unique perspective, and the way we see it determines how much we learn about certain issues, our level of concern about those issues, and the type of action (if any) that we will take to address the issues.

To determine how each of us views that relationship and the consequences of that view, we need to examine:

- the continuum between the Frontier World-View and the Sustainable World-View
- the Frontier World-View; and
- the Sustainable World-View

WORLD-VIEW CONTINUUM

The Frontier World-View is human-centered. In this view, the Earth is seen as a limitless and inexhaustible system within which humans are individuals, separate from each other and separate from nature. People holding this view do not believe that it is their responsibility to care for the earth, but rather that the earth and its resources exist to support them.

The Sustainable World-View is Earth-centered. In this view, the Earth is seen as a limited, but sustainable system-one which is able to survive for a specified period of time if its population and resources are managed so as to satisfy present needs without jeopardizing the future. People holding to this view see humans as connected to each other and as a part of nature.

In reality, most people do not hold strictly to either of these view-points. The world-view of any individual probably lies along a continuum between these polar view-points.

FRONTIER WORLD-VIEW

Human population growth is not considered a serious problem, and in fact is seen by some as being necessary for economic well-being. The argument is that more people will provide more brain power, which in turn will find solutions to environmental problems.

In this world-view, using energy from nonrenewable fossil fuel resources makes sense because it is less expensive to produce than its alternatives. By the time these resources have been depleted, we will have used the money we saved to develop substitutes. Increasing energy use is promoted, and the waste of energy is condoned, if this leads to greater economic growth and stability. The belief is that a strong economy will find ways to deal with the resulting waste products and increasing pollution.

Similarly, the use (and potential waste) of other nonrenewable resources is not seen as a problem, since prosperity will lead to the ability to solve the resulting problems.

The gap in life-sustaining resources (soil, food, water and air) between the "have" and "have not" nations is seen in this world-view as a fact of life, almost a "Darwinian" process which favors the strong and fit over the weak and less able. It is not seen as "natural" or productive for those who have been successful to sacrifice in order to aid those who have been less successful. Those who have more resources have "earned" them and it is not fair to expect these people to redistribute what they have to those who have not "earned" as much.

Finally, people holding to this world-view tend to believe that humans should not necessarily be obligated to protect and preserve the habitat of other living things. They believe that any successful form of life will naturally come to dominate its surroundings. Humans are currently one of the most successful species on earth and we have the right to take what we need to protect ourselves and make ourselves more conformable. This is exactly the way the members of any other species would act if they could.

SUSTAINABLE WORLD-VIEW

The Sustainable World-View sees Earth as a limited, but sustainable system. The bottom line in this world-view is that virtually anything that humans do to prolong the life of the planet, protect living things on the planet, and extend natural resources can be justified, even if it is not economically beneficial.

In this world-view, human population growth is considered a serious problem, one that is at the root of most environmental problems. People adhering to this view believe that the growth of population must be reduced until it reaches a sustainable level.

In this world-view, using energy from nonrenewable fossil fuel resources makes no sense at all, even though it is less expensive to produce than its alternatives. Moving toward the use of renewable or sustainable energy resources (generally those that are driven by the Sun) and reducing the consumption of energy will result in long-term availability and we will not need to find substitutes. Rather than trying to find ways to control pollution and other waste products, it makes more sense to find sources of energy which do not produce them.

Similarly, the use (and waste) of other nonrenewable resources is seen as a problem, because there might well be some resources for which we will not be able to find a substitute. The use of some nonrenewable resources is unavoidable, but in this world-view those who use these resources must strive to conserve (reduce consumption), reuse and recycle them and thus extend their availability.

The gap in life-sustaining resources (soil, food, water and air) between the "have" and "have not" nations is seen in this world-view to be potentially destabilizing and inhumane. Increased malnutrition and hunger in the "have not" nations must not be tolerated. In this world-view, those who consume life-sustaining resources must focus on ways to restore the quality of soil, water and air. They should strive to increase food production, to decrease the waste of food, and to find ways to redistribute food and water where necessary.

Finally, people holding to this world-view tend to believe that the destruction and elimination of habitats of any living things is not justifiable. We have already brought about the extinction of more species in the last 3 1/2 million years, than occurred in all of pre-human time. The resulting decline in biodiversity makes many living things more susceptible to disease and predators.

Those holding a Sustainable World-View believe that humans must see themselves as part of nature, not separate from it. The only thing that makes us special is our ability to alter marginally habitable environments and make them habitable. We are but one form of life in the long evolution of life. We must begin to act on the belief that the fate of the Planet is more important than the fate of humans. We are not yet at "the point of no return" for most major issues. However, we are at a critical time, perhaps the most critical time in the history of the Earth. We must act aggressively and responsibly, and we must begin now, if we want our children and our grandchildren to enjoy the quality of life that we have enjoyed. We have the ability to produce permanent, destructive alterations to the Earth. With that comes the responsibility not to do so.

HUMAN HEALTH

Public health problems caused by environmental contamination and emerging infectious diseases are a growing concern worldwide. These public health threats are affected by the relationship between people and the physical, chemical, and biological nature of our natural environments. Population growth and the associated pressures of development are increasing the difficulties associated with sustaining effective public health practices and policies. Vector-borne and zoonotic diseases, water contamination, airborne contaminants, bioaccumulative contaminants in the food chain, and environmental threats to public health the world over require marshalling of all our scientific knowledge and know-how to develop new solutions.

Understanding environmental and ecological health is a prerequisite to protecting public health. As the Nation's natural science agency, USGS can play a significant role in providing scientific knowledge and information that will improve our understanding of the environmental contributions to disease and human health.

Potential Contaminants and Pathogens in Air, Dusts and Soils

A variety of human health problems have been linked to exposures to dusts, other atmospheric aerosols, and soils. For example, inhalation exposures to asbestos, silica, and some metal-rich dusts in industrial or occupational settings have been recognized for decades as triggers for disease. Health concerns also arise from exposures to other anthropogenic atmospheric particulates, such as automobile exhaust, urban air pollution, smelting and coal combustion byproducts, and debris from disasters such as the World Trade Center collapse. Increasingly, environmental exposures to dusts and other atmospheric particulates (such as naturally-occurring asbestos, silica, volcanic ash, volcanic gas condensates, wildfire smoke, and dusts containing pathogens) are also being recognized as potential health concerns. Soils are recognized for their potential to affect human health, both as sources for deleterious dusts and pathogens, and, where contaminated by human activities, as sources for toxicity via ingestion exposures. A variety of USGS research activities are working to understand better the links between these geologic materials and human health.

Asbestiform Minerals and Human Health

Asbestos is a commercial-industrial term with a long history, and is not a mineralogical definition. "Asbestos" in the latter half of the 20th century became widely used in regulatory language to refer to well developed, long, thin particles (fibers or fibrils) and fiber bundles of specific mineral compositions, which have fulfilled particular industrial applications. In the U.S., asbestos is most commonly defined as the asbestiform variety of 6 naturally occurring hydrated silicate minerals; these include chrysotile, the asbestiform member of the serpentine group, and five minerals of the amphibole group:

(1) crocidolite, the asbestiform variety of riebeckite),

(2) amosite, the asbestiform variety of cummingtonite-grunerite,

(3) anthophyllite asbestos,

(4) actinolite asbestos, and

(5) tremolite asbestos.

According to the US EPA and the Agency for Toxic Substances and Disease Registry potential health effects of exposure to asbestos can include asbestosis, mesothelioma, and lung cancers. In the past, most medical research and regulatory focus has been on the health effects of asbestos used industrially or commercially. However, there have been increasing concerns recently regarding the potential health effects of naturally-occurring asbestos (NOA), which occurs as an accessory mineral in some rocks (for example, tremolite in metamorphosed carbonate rocks, chrysotile in serpentinite rocks), or as an impurity in some other industrial minerals (for example, amphibole asbestos intergrown with vermiculite mined at Libby, Montana).

Current USGS research activities in collaboration with health scientists examine a number of still unresolved scientific questions regarding asbestos, such as:

- Providing mineralogical information that will help health scientists better understand the potential role of fibrous but nonasbestiform crystals and cleavage fragments in disease;
- Compiling information on the range of geologic environments in which asbestiform minerals occur nationwide and globally;
- Developing remote sensing techniques to help map the distribution of potential naturally-occuring asbestos in rocks;
- Developing remote sensing techniques to differentiate asbestiform versus non-asbestiform varieties of the same mineral.
- Understanding how the mineralogical, chemical, and toxicological characteristics of the asbestos may vary between the different geologic environments of formation;
- Understanding how asbestiform minerals may interact chemically with lung, tissue, and cellular fluids, and how these interactions may influence toxicity.

4

Earth Materials and Human Health

THE BACKDROP

Earth materials generated by both natural and anthropogenic processes have been increasingly recognized as potential sources of illness in humans. For example, inhalation exposures to asbestos, silica, and some metal-rich dusts in industrial or occupational settings have been recognized for decades as triggers for certain diseases.

Health concerns also arise from exposures to other anthropogenic atmospheric particulates, such as automobile exhaust, urban air pollution, smelting and coal combustion byproducts, and debris from building demolition and incidents such as the World Trade Center collapse.

Increasingly, environmental exposures to dusts and other atmospheric particulates (such as naturally-occurring asbestos, silica, volcanic ash, volcanic gas condensates, wildfire smoke, and dusts containing pathogens) are also being recognized as potential health concerns. Soils are recognized for their potential to affect human health-as sources for dusts, sources for pathogens, and as sources for toxicity via any contained toxicants such as heavy metals or organic chemicals.

The Earth Materials and Human Health Project (EMHHP), in collaboration with medical and public health experts, is developing and applying a variety of earth science methods to help interpret the geologic links between earth materials and human health problems. The project's scientific results will provide a sound earth science basis for more informed, effective regulatory policies and remediation strategies.

The project is currently (fiscal years 2005-2009) funded primarily by the USGS Mineral Resources Program (MRP), with some additional funding from the Earth Surface Dynamics Program. The project is a follow-up of the MRP-funded Mineral Dusts and Human Health Project (MDHHP), which ran from fiscal years 2001-2004; key MDHHP activities, outcomes and publications are presented in a separate web site.

This project involves extensive collaboration with other USGS personnel and projects, and also health experts in a number of Federal and state agencies. Examples of other USGS projects with which this project is collaborating include the Near-Surface Processes, Industrial Minerals, Southwest Climate Change, Transoceanic Dusts, Surveys and Analysis, Geochemical Landscapes, and Wildfire Effects projects. Health experts with whom the project is collaborating include scientists from USEPA, NIOSH, OSHA, CDC, ATSDR, Public Health Service, Colorado Department of Public Health and Environment, University of Arizona, University of Colorado, SUNY Stonybrook, and other agencies and universities.

Mineral Dusts and Human Health

The Mineral Dusts and Human Health Project (MDHHP), which ran from fiscal year 2001 through fiscal year 2004, utilized an interdisciplinary approach (involving mineralogy, economic geology, aqueous and stable isotope geochemistry, analytical chemistry, remote sensing, regional geology, and toxicology expertise) to help understand how the geologic characteristics of mineral dusts (and the source materials from which the dusts are derived) may influence their roles in human health. A key aspect of the project was its integration of earth science and health science expertise and activities. A summary of MDHHP outcomes and publications is included below.

The project's primary focus was upon asbestos and fibrous dusts related to mining, mineral processing, or mineral products, and so the project addressed environmental and human health priorities of the Mineral Resources Program outlined in USGS science planning documents. To a limited extent, the project also successfully applied the same interdisciplinary approach (in collaboration as appropriate with other USGS projects) to study potential health implications of other earth materials such as: metal-bearing mine wastes, mill tailings, and smelter emissions; dusts from dry lake beds; soils; volcanic ash; coal and coal fly ash; and dusts from building collapse. The studies identified many topics for a spectrum of earth materials where substantial further research is needed to address increasing societal concerns.

The majority of the project's funding came from the USGS Mineral Resources Program. The Mendenhall Program supported a post-doctoral toxicology position linked to the project for two years, and the Earth Surface Dynamics Program and Energy Resources program contributed support for the toxicology position.

Background

The Mineral Dusts and Human Health project's focus was primarily on asbestos-containing dusts, although other dusts and dust sources were also investigated to a lesser extent. The project was initiated to provide impartial scientific input to help address renewed societal and regulatory concerns about potential health effects associated with exposure to asbestos. In the past, societal and regulatory concerns were focused on commercial and industrial asbestos. However, in recent years, largely as a result of significant health problems at Libby, Montana, concerns have increased substantially regarding so-called "naturally-occurring asbestos" (NOA) and other fibrous minerals that occur a) as accessory minerals in other industrial mineral deposits (such as vermiculite deposits like those mined at Libby, and such as some talc deposits) and b) in rock units (such as serpentinite-bearing ultramafic rocks).

The project provided insights about a number of asbestos issues that benefit a wide variety of stakeholders. The project's activities also provided information that can be used to help address some of the many unanswered questions remaining about asbestos. For example, many questions still remain about how asbestos actually causes toxicity, and whether or not fibrous but non-asbestiform varieties of the same minerals can also trigger toxicity. Further, relatively little is known about the full range of geologic environments in which asbestos or other fibrous minerals can occur, the extent to which natural erosion or anthropogenic disturbance of these sources contributes to background levels of asbestos in the air, and the extent to which such background contributions can themselves trigger disease. Although the project's primary focus remained on asbestos and related minerals, limited studies were also carried out on other mineral particulates and/or their potential source material, including: mine wastes; mill tailings; soils affected by smelter emissions; mercury mine calcines; dry lake beds such as Owens Lake, CA; volcanic ash; various soils; coal dust and coal fly ash; and dusts generated by collapse of buildings such as the World Trade Center. These studies demonstrated that a similar interdisciplinary approach to that developed by the project to study asbestos can provide important insights into potential health concerns tied to many other types of earth materials that may be liberated as particulates into the environment. The studies also identified many topics in this realm where substantial further research is needed to address increasing societal concerns. The USGS Mineral Resources Program is currently (FY 2005-2009) funding a follow-up project titled "Earth Materials and Human Health" to address unresolved questions about asbestos and the potential health impacts of other earth materials such as mine wastes, soils, volcanic ash, dry lake bed dusts, coal dust and coal fly ash, and others.

MINERAL DUSTS AND HUMAN HEALTH (MDHH) PROJECT OUTCOMES

Asbestos

- USGS characterization work on fibrous amphiboles and vermiculite at Libby, Montana (funded in part by the US EPA and in part by the USGS Mineral Resources Program) has provided key scientific information that is being used or will be used to help:

  a. Assess the nature and extent of amphibole contamination present at Libby, at hundreds of plants nationwide where Libby vermiculite was processed, and in approximately a million homes with vermiculite insulation from Libby.

b. Guide remedial efforts at sites where fibrous amphiboles from Libby are present.

c. Understand how the fibrous amphiboles at Libby and geologically similar deposit types influence toxicity

Results of the USGS activities will directly support or influence vermiculite-related cleanup activities nationwide.

- The project's geologic studies of vermiculite deposits nationwide:

    a. Indicate that not all vermiculite deposits contain fibrous amphiboles, and therefore not all types of vermiculite deposits need to be regulated as though they contain asbestos.

    b. Provide methods that are being used to help assess whether a particular vermiculite sample is from deposit types likely to contain fibrous amphiboles.

- The project's studies of the geologic occurrences of asbestiform or other potentially toxic fibrous minerals:

    a. Provide a geologic model of formation that helps explain why some types of talc deposits do not contain asbestiform amphiboles and why others do.

    b. Demonstrate that there are many possible geologic sources for asbestiform or other potentially toxic fibrous minerals nationwide. These sources, through either natural erosion or anthropogenic disturbance, may contribute to background levels of fibrous particulates in the air, and so must be taken into account in the interpretation of epidemiological data on asbestos-related diseases, and in the development of appropriate air quality standards for asbestos.

- The project systematically compared the mineralogical, geochemical, and toxicological properties of a wide variety of asbestos-related toxicological standards. Results show that there are considerable variations in all of these properties between different standards of a given asbestos mineral. Toxicological studies do not routinely take such variations into account, which may help explain seemingly conflicting results of different studies using different standards of the same minerals.

- The project developed, tested, and demonstrated the utility of AVIRIS (Airborne Visible and Infra-Red Imaging Spectrometer) remote sensing techniques to map the occurrence of potentially asbestos-forming minerals over large areas. The technique is particularly valuable for helping evaluate areas where potentially asbestos bearing rock units may occur, but where geologic mapping is limited. This activity was carried out in collaboration with colleagues at the California Geological Survey.
- As a result of these project activities, USGS has responded to many requests for expert geologic and mineralogic information on asbestos, including:
    a. A request from the American Thoracic Society to write a summary of asbestos mineralogy for its revised criteria for assessing asbestos-related disease.
    b. A request to participate in the Federal Interagency Working Group on Asbestos. This Working Group is charged with evaluating current asbestos-related issues, and how these issues can be addressed by the regulatory agencies based on sound science input from science agencies such as the USGS. Regular participation by project members in this working group highlights the need for the USGS' impartial earth science input into asbestos issues.
    c. Multiple inquiries by other Federal agencies, State agencies, industry, and the public for scientific information on asbestos-related issues.
    d. A formal request for a project scientist to serve as an expert witness in Federal litigation related to Libby.
    e. A formal request for a project scientist to serve as an expert member of a committee overseeing activities of the Congressionally mandated U.S. Navy Lung Disease Assessment Program.

World Trade Center Dust Characterization

- Based on the characterization work done by USGS-MDHHP members on Libby, the USEPA and US Public Health Service, in the days immediately following 9-11, requested USGS assistance to assess the amounts and spatial distribution of asbestos in the dusts deposited by the collapse of the World Trade Center (WTC) towers. This emergency response effort was carried out by MDHHP scientists, was funded under the auspices of the MDHHP, and utilized the full mix of analytical capabilities developed by the project to study asbestos and other dust-related health issues.
- A field crew was mobilized to lower Manhattan area by September 16, collected more than 35 samples of the dusts from around lower Manhattan, and returned the samples to the Denver labs for analysis by September 20.
- AVIRIS thermal images of Ground Zero showing locations of burning "hot spots" in the debris were provided to emergency responders by September 18, for use in fighting fires on site.
- Preliminary USGS findings on the dusts were released on September 27, 2001, to emergency response authorities. The USGS studies provided an early and detailed (in terms of types of analyses and numbers of samples analyzed) summary of the dust mineralogy, chemical composition, and geochemical reactivity.
- A key USGS result was the conclusion that amphibole asbestos was likely not present in anything but very low levels in the dusts. However, the USGS results did indicate the presence of chrysotile asbestos in levels around 1-2%.
- The USGS study provided an early measure of the chemically reactive, alkaline nature of the dusts. The study also provided insights into the processes by which the dusts may have interacted chemically with water.
- USGS results have been cited by many different government agencies, Congressional representatives, the

media, and the public. Project scientists continue to receive requests for information and results nearly 4 years later, and continue to field invitations to give presentations on the results.

- The study demonstrated that there is an appropriate role for a natural science agency such as the USGS in emergency response situations involving earth materials.
- The study and other related project work has resulted in a formal request for a project scientist to serve as a member of the World Trade Center Expert Technical Review Panel established by the EPA in consultation with the White House Council on Environmental Quality.

The USGS World Trade Center Response would not have been possible without the types of expertise developed through many years of project activities funded by the Mineral Resources program.

Characterization of Non-asbestiform Dusts and their Sources

- USGS-MDHHP members, in cooperation with USGS scientists on other projects, have carried out reconnaissance studies of other atmospheric particulates and their sources. These studies show that the same interdisciplinary approach used for asbestos can successfully be applied to help understand how mineralogical and chemical characteristics of dusts and their source materials may influence human health.
- The project integrates mineralogical, geochemical reactivity, and toxicity characterization tests of earth materials to evaluate the role of particle mineralogy and reactivity in toxicity. For example, dusts from Owens Lake are well known for their high arsenic contents. Work in collaboration with the USGS Southwest Dusts project shows that the arsenic is likely to be quite bioaccessible, and that the dusts also contain other bioaccessible, potentially toxic, elements such as chromium.

- Early results of MDHHP studies characterizing volcanic ash have led to an invitation for a project scientist to serve as an expert member of the newly formed International Volcanic Health Hazard Network.

NATURAL ENVIRONMENT.

The natural environment comprises all living and non-living things that occur naturally on Earth. In its purest sense, it is thus an environment that is not the result of human activity or intervention. The natural environment may be contrasted to "the built environment", and is also in contrast to the concept of cultural landscape.

For some, there is a difficulty with the term "natural environment" in that nearly all environments have been directly or indirectly influenced by humans at some point in time. In order to address this concern, some level of human influence is thus allowable without the status of any particular landscape ceasing to be "natural." The term's meaning, however, is usually dependent more on context than a set definition. Many natural environments are the product of the interaction between nature and humans. For this reason, the term ecosystem has been used to describe an environment that contains nature, and includes people. It follows then that environmental problems are human or social problems. Some also consider it dangerously misleading to regard "environment" as separate from "people."

CHALLENGES

It is the common understanding of natural environment that underlies environmentalism-a broad political, social, and philosophical movement that advocates various actions and policies in the interest of protecting what nature remains in the natural environment, or restoring or expanding the role of nature in this environment. While wilderness is increasingly rare, wild nature (e.g., unmanaged forests, uncultivated grasslands, wildlife, wildflowers) can be found in many locations previously inhabited by humans.

Goals commonly expressed by the environmentalists include: reduction and clean up of man-made pollution, with future goals of zero pollution; reducing societal consumption of non-renewable fuels, development of alternative, green, low carbon or renewable energy sources; conservation and sustainable use of scarce resources such as water, land and air; protection of representative or unique or pristine ecosystems; preservation and expansion of threatened or endangered species or ecosystems from extinction; the establishment of nature and biosphere reserves under various types of protection, and, most generally, the protection of biodiversity and ecosystems upon which all human and other life on earth depends.

More recently, there has been a strong concern about climatic changes caused by anthroprogenic releases of greenhouse gases, most notably carbon dioxide, and their interactions with human uses and the natural environment. Efforts here have focused on the mitigation of greenhouse gases that are causing climatic changes (i.e., through the Climate Change Convention and the Kyoto Protocol), and on developing adaptative strategies to assist species, ecosystems, humans, regions and nations in adjusting to these climatic changes.

A more profound challenge however, is to identify the natural environmental dynamics in contrast to environmental changes not within natural variances. A common solution is to adapt a static view neglecting natural variances to exist. Methodologically this view could be defended when looking at processes which changes slowly and short time series, while the problem arrives when fast processes turns essential in the object of the study.

BUILT ENVIRONMENT

The phrase built environment refers to the manmade surroundings that provide the setting for human activity, ranging from the large-scale civic surroundings to the personal places.

In architecture and environmental psychology, the phrase is a useful acknowledgment that a small fraction of buildings constructed annually, even in the industrialized world, are designed by architects, and that users of the built environment encounter issues that cross the traditional professional boundaries between urban planners, traffic engineers, zoning authorities, architects, interior designers, industrial designers, etc. Historically, much of the built environment has taken the form of vernacular architecture, and this is still the case in large parts of the world. In the industrialized world, many buildings are produced by large scale development remote from its eventual users.

In landscape architecture, the built environment is identified as opposed to the natural environment, with the recognition that places like Central Park may have the look, feel, and nourishing quality of natural surroundings while being completely artificial and "built", thus blurring the line between the two.

In urban planning, the phrase connotes the idea that a large percentage of the human environment is manmade, and these artificial surroundings are so extensive and cohesive that they function as organisms in the consumption of resources, disposal of wastes, and facilitation of productive enterprise within its bounds. Recently there has also been considerable dialogue and research into the impact of the built environment's impact on population health.

CULTURAL LANDSCAPE

The Dresden Elbe Valley World Heritage Site is according to the UNESCO "an outstanding example of land use, representing an exceptional development of a major Central-European city" having almost half a million inhabitants.

Cultural landscape is defined as the human-modified environment, including fields, houses, churches, highways, planted forests, and mines, as well as weeds and pollution.

A cultural landscape defined as: "A geographic area, including both cultural and natural resources and the wildlife or domestic animals therein, associated with an historic event, activity, or person or exhibiting other cultural or aesthetic values."

In the USA, there are four general types of cultural landscapes, not mutually exclusive: historic sites, historic designed landscapes, historic vernacular landscapes, and ethnographic landscapes.

"The cultural landscape is fashioned from a natural landscape by a cultural group. Culture is the agent, the natural are the medium, the cultural landscape is the result. Under the influence of a given culture, itself changing through time, the landscape undergoes development, passing through phases and probably reaching ultimately the end of its cycle of development. With the introduction of a different, alien culture, a rejuvenation of the cultural landscape sets in, or a new landscape is superimposed on remnants of the old one" (Sauer's, 1925).

Sauer was explicitly concerned to counter an environmental determinism which had dominated the American geography of the previous generation, within which human agency was given scant autonomy in the shaping of the visible landscape. Sauer believed, and was determined to stress the agency of culture as a force in shaping the visible features of the Earth's surface in delimited areas. Within his definition, the physical environment retains a central significance, as the medium with and through which human cultures act. Interestingly this results in elements of the physical environment, such as topography, soils, plants and animals needing to be incorporated into studies of the cultural landscape; so far as they provoke human responses and adaptations, or have themselves been altered by human activity (e.g. forest clearing and dams). Cronon (1995) believes that Sauer's definition cannot be upheld today since it is clear that there is no clear distinction between nature and culture, since both interlink and should be regarded together as co-productions.

The Myth of the Pristine Landscape

With the arrival of Europeans to North and South America and their subsequent forays into the hearts of these continents, explorers frequently encountered sparsely inhabited landscapes they thought was untouched wilderness, a belief which has persisted into contemporary time. In a 1992 journal article, William M. Denevan closely cxamines this idea of "a world of barely perceptible human disturbance" and instead argues for the alternative hypothesis in which the state of American landscapes as they were in early post-contact times were largely shaped by anthropogenic processes such as deforestation and agricultural burning. These extensive modifications done on the part of pre-contact peoples to their local ecosystems effectively transformed them into cultural and humanized landscapes. That indigenous peoples were capable of altering the landscapes in which they lived assigns much more agency to them than was previously credited to them for the past several centuries. It also refutes the notion of the 'noble savage' who lived in immaculate harmony with nature and left little to no impact on his landscape.

ECOLOGY

Ecology, œcology or ecological science, is the scientific study of the distribution and abundance of living organisms and how the distribution and abundance are affected by interactions between the organisms and their environment. The environment of an organism includes both physical properties, which can be described as the sum of local abiotic factors such as insolation (sunlight), climate, and geology, as well as the other organisms that share its habitat. The term oekologie was coined in 1866 by the German biologist Ernst Haeckel the word is derived from the Greek and therefore "ecology" means the "study of the household [of nature]".

The word "ecology" is often used in common parlance as a synonym for the natural environment or environmentalism. Likewise "ecologic" or "ecological" is often taken in the sense of environmentally friendly.

Scope

Ecology is usually considered a branch of biology, the general science that studies living organisms. Organisms can be studied at many different levels, from proteins and nucleic acids (in biochemistry and molecular biology), to cells (in cellular biology), to individuals (in botany, zoology, and other similar disciplines), and finally at the level of populations, communities, and ecosystems, to the biosphere as a whole; these latter strata are the primary subjects of ecological inquiries. Ecology is a multi-disciplinary science. Because of its focus on the higher levels of the organization of life on earth and on the interrelations between organisms and their environment, ecology draws heavily on many other branches of science, especially geology and geography, meteorology, pedology, genetics, chemistry, and physics. Thus, ecology is considered by some to be a holistic science, one that over-arches older disciplines such as biology which in this view become sub-disciplines contributing to ecological knowledge.

Agriculture, fisheries, forestry, medicine and urban development are among human activities that would fall within Krebs' (1972: 4) explanation of his definition of ecology: "where organisms are found, how many occur there, and why".

As a scientific discipline, ecology does not dictate what is "right" or "wrong". However, ecological knowledge such as the quantification of biodiversity and population dynamics have provided a scientific basis for expressing the aims of environmentalism and evaluating its goals and policies. Additionally, a holistic view of nature is stressed in both ecology and environmentalism. Consider the ways an ecologist might approach studying the life of honeybees:

- The behavioral relationship between individuals of a species is behavioral ecology-for example, the study of the queen bee, and how she relates to the worker bees and the drones.
- The organized activity of a species is community ecology; for example, the activity of bees assures the pollination of flowering plants. Bee hives additionally produce honey which is consumed by still other species, such as bears.
- The relationship between the environment and a species is environmental ecology-for example, the consequences of environmental change on bee activity. Bees may die out due to environmental changes. The environment simultaneously affects and is a consequence of this activity and is thus intertwined with the survival of the species.

Disciplines of Ecology

Ecology is a broad discipline comprised of many sub-disciplines. A common, broad classification, moving from lowest to highest complexity, where complexity is defined as the number of entities and processes in the system under study, is:

- Ecophysiology and Behavioral ecology examine adaptations of the individual to its environment.
- Autecology studies the dynamics of populations of a single species.
- Community ecology (or synecology) focuses on the interactions between species within an ecological community.
- Ecosystem ecology studies the flows of energy and matter through the biotic and abiotic components of ecosystems.
- Landscape ecology examines processes and relationship across multiple ecosystems or very large geographic areas.

Ecology can also be sub-divided according to the species of interest into fields such as animal ecology, plant ecology, insect ecology, and so on. Another frequent method of subdivision is by biome studied, e.g., Arctic ecology (or polar ecology), tropical ecology, desert ecology, etc. The primary technique used for investigation is often used to subdivide the discipline into groups such as chemical ecology, genetic ecology, field ecology, statistical ecology, theoretical ecology, and so forth. Note that these different systems are unrelated and often applied at the same time; one could be a theoretical plant community ecologist, or a polar ecologist interested in animal genetics.

5

Biosphere

INTRODUCTION

For modern ecologists, ecology can be studied at several levels: population level (individuals of the same species in the same or similar environment), biocoenosis level (or community of species), ecosystem level, and biosphere level.

The outer layer of the planet Earth can be divided into several compartments: the hydrosphere (or sphere of water), the lithosphere (or sphere of soils and rocks), and the atmosphere (or sphere of the air).

The biosphere (or sphere of life), sometimes described as "the fourth envelope", is all living matter on the planet or that portion of the planet occupied by life. It reaches well into the other three spheres, although there are no permanent inhabitants of the atmosphere. Relative to the volume of the Earth, the biosphere is only the very thin surface layer which extends from 11,000 meters below sea level to 15,000 meters above.

It is thought that life first developed in the hydrosphere, at shallow depths, in the photic zone. (Although recently a competing theory has emerged, that life originated around hydrothermal vents in the deeper ocean) Multicellular organisms then appeared and colonized benthic zones.

Photosynthetic organisms gradually produced the chemically unstable oxygen-rich atmosphere that characterizes our planet. Terrestrial life developed later, after the ozone layer protecting living beings from UV rays formed. Diversification of terrestrial species is thought to be increased by the continents drifting apart, or alternately, colliding.

Biodiversity is expressed at the ecological level (ecosystem), population level (intraspecific diversity), species level (specific diversity), and genetic level. Recently technology has allowed the discovery of the deep ocean vent communities. This remarkable ecological system is not dependent on sunlight but bacteria, utilising the chemistry of the hot volcanic vents, are at the base of its food chain.

The biosphere contains great quantities of elements such as carbon, nitrogen and oxygen. Other elements, such as phosphorus, calcium, and potassium, are also essential to life, yet are present in smaller amounts. At the ecosystem and biosphere levels, there is a continual recycling of all these elements, which alternate between the mineral and organic states.

While there is a slight input of geothermal energy, the bulk of the functioning of the ecosystem is based on the input of solar energy. Plants and photosynthetic microorganisms convert light into chemical energy by the process of photosynthesis, which creates glucose (a simple sugar) and releases free oxygen.

Glucose thus becomes the secondary energy source which drives the ecosystem. Some of this glucose is used directly by other organisms for energy. Other sugar molecules can be converted to other molecules such as amino acids. Plants use some of this sugar, concentrated in nectar to entice pollinators to aid them in reproduction.

Cellular respiration is the process by which organisms (like mammals) break the glucose back down into its constituents, water and carbon dioxide, thus regaining the stored energy the sun originally gave to the plants. The proportion of photosynthetic activity of plants and other photosynthesizers to the respiration of other organisms determines the specific composition of the Earth's atmosphere, particularly its oxygen level. Global air currents mix the atmosphere and maintain nearly the same balance of elements in areas of intense biological activity and areas of slight biological activity.

Water is also exchanged between the hydrosphere, lithosphere, atmosphere and biosphere in regular cycles. The oceans are large tanks, which store water, ensure thermal and climatic stability, as well as the transport of chemical elements thanks to large oceanic currents.

For a better understanding of how the biosphere works, and various dysfunctions related to human activity, American scientists simulated the biosphere in a small-scale model, called Biosphere II.

6

Ecosystem

The first principle of ecology is that each living organism has an ongoing and continual relationship with every other element that makes up its environment. An ecosystem can be defined as any situation where there is interaction between organisms and their environment.

The ecosystem is composed of two entities, the entirety of life, the biocoenosis and the medium that life exists in, the biotope. Within the ecosystem, species are connected by food chains or food webs. Energy from the sun, captured by primary producers via photosynthesis, flows upward through the chain to primary consumers (herbivores), and then to secondary and tertiary consumers (carnivores), before ultimately being lost to the system as waste heat. In the process, matter is incorporated into living organisms, which return their nutrients to the system via decomposition, forming biogeochemical cycles such as the carbon and nitrogen cycles.

CONCEPT OF ECOSYSTEM

The concept of an ecosystem can apply to units of variable size, such as a pond, a field, or a piece of deadwood. A unit of smaller size is called a microecosystem. For example, an ecosystem can be a stone and all the life under it. A mesoecosystem could be a forest, and a macroecosystem a whole ecoregion, with its drainage basin.

The main questions when studying an ecosystem are:

- Whether the colonization of a barren area could be carried out

- Investigation the ecosystem's dynamics and changes
- The methods of which an ecosystem interacts at local, regional and global scale
- Whether the current state is stable
- Investigating the value of an ecosystem and the ways and means that interaction of ecological systems provide benefit to humans, especially in the provision of healthy water.

Ecosystems are often classified by reference to the biotopes concerned. The following ecosystems may be defined:

- As continental ecosystems, such as forest ecosystems, meadow ecosystems such as steppes or savannas), or agro-ecosystems
- As ecosystems of inland waters, such as lentic ecosystems such as lakes or ponds; or lotic ecosystems such as rivers
- As oceanic ecosystems.

Another classification can be done by reference to its communities, such as in the case of an human ecosystem.

DYNAMICS AND STABILITY

Ecological factors which affect dynamic change in a population or species in a given ecology or environment are usually divided into two groups: abiotic and biotic.

Abiotic factors are geological, geographical, hydrological and climatological parameters. A biotope is an environmentally uniform region characterized by a particular set of abiotic ecological factors. Specific abiotic factors include:

- Water, which is at the same time an essential element to life and a milieu
- Air, which provides oxygen, nitrogen, and carbon dioxide to living species and allows the dissemination of pollen and spores
- Soil, at the same time source of nutriment and physical support

  - o Soil pH, salinity, nitrogen and phosphorus content, ability to retain water, and density are all influential
- Temperature, which should not exceed certain extremes, even if tolerance to heat is significant for some species
- Light, which provides energy to the ecosystem through photosynthesis
- Natural disasters can also be considered abiotic

Biocenose, or community, is a group of populations of plants, animals, micro-organisms. Each population is the result of procreations between individuals of same species and cohabitation in a given place and for a given time. When a population consists of an insufficient number of individuals, that population is threatened with extinction; the extinction of a species can approach when all biocenoses composed of individuals of the species are in decline. In small populations, consanguinity (inbreeding) can result in reduced genetic diversity that can further weaken the biocenose.

Biotic ecological factors also influence biocenose viability; these factors are considered as either intraspecific and interspecific relations.

Intraspecific relations are those which are established between individuals of the same species, forming a population. They are relations of cooperation or competition, with division of the territory, and sometimes organization in hierarchical societies.

An ant lion lies in wait under its pit trap, built in dry dust under a building, awaiting unwary insects that falls in. Many pest insects are partly or wholly controlled by other insect predators.

Interspecific relations-interactions between different species-are numerous, and usually described according to their beneficial, detrimental or neutral effect (for example, mutualism (relation ++) or competition (relation--). The most significant relation is the relation of predation (to eat or to be eaten), which leads to the essential concepts in ecology of food chains (for example, the grass is consumed by the herbivore, itself consumed by a carnivore, itself consumed by a carnivore of larger size). A high predator to prey ratio can have a negative influence on both the predator and prey biocenoses in that low availability of food and high death rate prior to sexual maturity can decrease (or prevent the increase-of) populations of each, respectively. Selective hunting of species by humans which leads to population decline is one example of a high predator to prey ratio in action. Other interspecific relations include parasitism, infectious disease and competition for limiting resources, which can occur when two species share the same ecological niche.

The existing interactions between the various living beings go along with a permanent mixing of mineral and organic substances, absorbed by organisms for their growth, their maintenance and their reproduction, to be finally rejected as waste. These permanent recyclings of the elements (in particular carbon, oxygen and nitrogen) as well as the water are called biogeochemical cycles. They guarantee a durable stability of the biosphere (at least when unchecked human influence and extreme weather or geological phenomena are left aside). This self-regulation, supported by negative feedback controls, ensures the perenniality of the ecosystems. It is shown by the very stable concentrations of most elements of each compartment. This is referred to as homeostasis. The ecosystem also tends to evolve to a state of ideal balance, reached after a succession of events, the climax (for example a pond can become a peat bog).

Spatial Relationships and Subdivisions of Land

Ecosystems are not isolated from each other, but are interrelated. For example, water may circulate between ecosystems by the means of a river or ocean current. Water itself, as a liquid medium, even defines ecosystems. Some species, such as salmon or freshwater eels move between marine systems and fresh-water systems. These relationships between the ecosystems lead to the concept of a biome.

Biome: A biome is a homogeneous ecological formation that exists over a large region as tundra or steppes. The biosphere comprises all of the Earth's biomes--the entirety of places where life is possible--from the highest mountains to the depths of the oceans. Biomes correspond rather well to subdivisions distributed along the latitudes, from the equator towards the poles, with differences based on to the physical environment (for example, oceans or mountain ranges) and to the climate. Their variation is generally related to the distribution of species according to their ability to tolerate temperature and/or dryness. For example, one may find photosynthetic algae only in the photic part of the ocean (where light penetrates), while conifers are mostly found in mountains.

Though this is a simplification of more complicated scheme, latitude and altitude approximate a good representation of the distribution of biodiversity within the biosphere. Very generally, the richness of biodiversity (as well for animal than plant species) is decreasing most rapidly near the equator and less rapidly as one approaches the poles.

The biosphere may also be divided into ecozone, which are very well defined today and primarily follow the continental borders. The ecozones are themselves divided into ecoregions, though there is not agreement on their limits.

ECOSYSTEM PRODUCTIVITY

In an ecosystem, the connections between species are generally related to food and their role in the food chain. There are three categories of organisms:

- Producers--usually plants which are capable of photosynthesis but could be other organisms such as bacteria around ocean vents that are capable of chemosynthesis.
- Consumers--animals, which can be primary consumers (herbivorous), or secondary or tertiary consumers (carnivorous).
- Decomposers--bacteria, mushrooms which degrade organic matter of all categories, and restore minerals to the environment.

These relations form sequences, in which each individual consumes the preceding one and is consumed by the one following, in what are called food chains or food network. In a food network, there will be fewer organisms at each level as one follows the links of the network up the chain.

These concepts lead to the idea of biomass (the total living matter in a given place), of primary productivity (the increase in the mass of plants during a given time) and of secondary productivity (the living matter produced by consumers and the decomposers in a given time).

These two last ideas are key, since they make it possible to evaluate the load capacity--the number of organisms which can be supported by a given ecosystem. In any food network, the energy contained in the level of the producers is not completely transferred to the consumers. And the higher one goes up the chain, the more energy and resources is lost and consumed.

Thus, from an energy-and environmental-point of view, it is more efficient for humans to be primary consumers (to subsist from vegetables, grains, legumes, fruit, cotton, etc.) than as secondary consumers (from eating herbivores, omnivores, or their products, such as milk, chickens, cattle, sheep, etc.) and still more so than as a tertiary consumer (from consuming carnivores, omnivores, or their products, such as fur, pigs, snakes, alligators, etc.). An ecosystem(s) is unstable when the load capacity is overrun and is especially unstable when a population doesn't have an ecological niche and overconsumers.

The productivity of ecosystems is sometimes estimated by comparing three types of land-based ecosystems and the total of aquatic ecosystems:

- The forests (1/3 of the Earth's land area) contain dense biomasses and are very productive. The total production of the world's forests corresponds to half of the primary production.
- Savannas, meadows, and marshes (1/3 of the Earth's land area) contain less dense biomasses, but are productive. These ecosystems represent the major part of what humans depend on for food.
- Extreme ecosystems in the areas with more extreme climates--deserts and semi-deserts, tundra, alpine

meadows, and steppes--(1/3 of the Earth's land area) have very sparse biomasses and low productivity

- Finally, the marine and fresh water ecosystems (3/4 of Earth's surface) contain very sparse biomasses (apart from the coastal zones).

Humanity's actions over the last few centuries have seriously reduced the amount of the Earth covered by forests (deforestation), and have increased agro-ecosystems (agriculture). In recent decades, an increase in the areas occupied by extreme ecosystems has occurred (desertification).

ECOLOGICAL CRISIS

Generally, an ecological crisis occurs with the loss of adaptive capacity when the resilience of an environment or of a species or a population evolves in a way unfavourable to coping with perturbations that interfere with that ecosystem, landscape or species survival.

It may be that the environment quality degrades compared to the species needs, after a change in an abiotic ecological factor (for example, an increase of temperature, less significant rainfalls). It may be that the environment becomes unfavourable for the survival of a species (or a population) due to an increased pressure of predation (for example overfishing).

Lastly, it may be that the situation becomes unfavourable to the quality of life of the species (or the population) due to a rise in the number of individuals (overpopulation).

Ecological crises may be more or less brutal (occurring within a few months or taking as long as a few million years). They can also be of natural or anthropic origin. They may relate to one unique species or to many species.

Lastly, an ecological crisis may be local (as an oil spill) or global (a rise in the sea level due to global warming).

According to its degree of endemism, a local crisis will have more or less significant consequences, from the death of many individuals to the total extinction of a species. Whatever its origin, disappearance of one or several species often will involve a rupture in the food chain, further impacting the survival of other species. In the case of a global crisis, the consequences can be much more significant; some extinction events showed the disappearance of more than 90% of existing species at that time. However, it should be noted that the disappearance of certain species, such as the dinosaurs, by freeing an ecological niche, allowed the development and the diversification of the mammals. An ecological crisis thus paradoxically favored biodiversity.

Sometimes, an ecological crisis can be a specific and reversible phenomenon at the ecosystem scale. But more generally, the crises impact will last. Indeed, it rather is a connected series of events, that occur till a final point. From this stage, no return to the previous stable state is possible, and a new stable state will be set up gradually.

Lastly, if an ecological crisis can cause extinction, it can also more simply reduce the quality of life of the remaining individuals. Thus, even if the diversity of the human population is sometimes considered threatened, few people envision human disappearance at short span. However, epidemic diseases, famines, impact on health of reduction of air quality, food crises, reduction of living space, accumulation of toxic or non degradable wastes, threats on keystone species (great apes, panda, whales) are also factors influencing the well-being of people.

During the past decades, this increasing responsibility of humanity in some ecological crises has been clearly observed. Due to the increases in technology and a rapidly increasing population, humans have more influence on their own environment than any other ecosystem engineer.

Some usually quoted examples as ecological crises are:

- Permian-Triassic extinction event 250 million of years ago
- Cretaceous-Tertiary extinction event 65 million years ago
- Global warming related to the Greenhouse effect. Warming could involve flooding of the Asian deltas, multiplication of extreme weather phenomena and changes in the nature and quantity of the food resources.

- Ozone layer hole issue
- Deforestation and desertification, with disappearance of many species.
- The nuclear meltdown at Chernobyl in 1986 caused the death of many people and animals from cancer, and caused mutations in a large number of animals and people. The area around the plant is now abandoned by humans because of the large amount of radiation generated by the meltdown. Twenty years after the accident, the animals have returned.

ENVIRONMENT

Environment refers to a complex of surrounding circumstances, conditions, or influences in which a thing is situated or is developed. Environment is the external forces affecting living things, while nature is the inner force.

It may also refer to:

- Natural environment, all living and non-living things that occur naturally on Earth
- Environmental science, the study of the interactions among the physical, chemical and biological components of the environment
- Environmentalism, a concern with the preservation of the natural environment
- Social environment, the culture that an individual lives in, and the people and institutions with whom they interact
- The set of environment variables defined in a process.
- Runtime environment, a virtual machine state which provides software services for processes or programs while a computer is running
- Integrated development environment, a type of computer software that assists computer programmers in developing software
- Desktop environment, in computing, is a graphical user interface to the computer.

7

Environmental Science

Environmental science is the study of the interactions among the physical, chemical and biological components of the environment; with a focus on pollution and degradation of the environment related to human activities; and the impact on biodiversity and sustainability from local and global development. It is inherently an interdisciplinary field that draws upon not only its core scientific areas, but also applies knowledge from other non-scientific studies such as economics, law and social sciences. Physics is used to understand the flux of material and energy interaction and construct mathematical models of environmental phenomena. Chemistry is applied to understand the molecular interactions in natural systems. Biology is fundamental to describing the effects within the plant and animal kingdoms.

A SUBSTANTIVE CONCEPT

While the concept of environmental science has existed for centuries, it came alive as a substantive, active field of scientific investigation in the 1960s and 1970s driven by (a) the need for a large multi-disciplined team to analyze complex environmental problems, (b) the arrival of substantive environmental laws requiring specific environmental protocols of investigation and (c) the growing public awareness of a need for action in addressing environmental problems.

Environmental science encompasses issues such as climate change, conservation, biodiversity, groundwater and soil contamination, use of natural resources, waste management, sustainable development, air pollution and noise pollution. Due to the inherent interdisciplinary nature of environmental science, teams of professionals commonly work together to conduct environmental research or to produce Environmental Impact Statements, as required by the U.S. National Environmental Policy Act (NEPA) or under state laws. There are professional organizations that engender work in environmental science and aid in communication among the diverse sciences.

SUB-CATEGORIES

Atmospheric sciences examines the phenomenology of the Earth's gaseous outer layer with emphasis upon interrelation to other systems. Atmospheric sciences comprises meteorological studies, greenhouse gas phenomena, atmospheric dispersion modeling of airborne contaminants, sound propagation phenomena related to noise pollution, and even light pollution

Taking the example of the global warming phenomena, physicists create computer models of atmospheric circulation and infra-red radiation transmission, chemists examine the inventory of atmospheric chemicals and their reactions, biologists analyze the plant and animal contributions to carbon dioxide fluxes, and specialists such as meteorologists and oceanographers add additional breadth in understanding the atmospheric dynamics.

Ecology studies typically analyze the dynamics among an interrelated set of populations, or a population and some aspect of its environment. These studies could address endangered species, predator/prey interactions, habitat integrity, effects upon populations by environmental contaminants, or impact analysis of proposed land development upon species viability.

An interdisciplinary analysis of an ecological system which is being impacted by one or more stressors might include several related environmental science fields. For example one might examine an estuarine setting where a proposed industrial development could impact certain species by water pollution and air pollution. For this study biologists would describe the flora and fauna, chemists would analyze the transport of water pollutants to the marsh, physicists would calculate air pollution emissions and geologists would assist in understanding the marsh soils and bay muds. Environmental chemistry is the study of chemical alterations in the environment. Principal areas of study include soil contamination and water pollution. The topics of analysis involve chemical degradation in the environment, multi-phase transport of chemicals (for example, evaporation of a solvent containing lake to yield solvent as an air pollutant), and chemical effects upon biota.

As an example study, consider the case of a leaking solvent tank which has entered the soil upgradient of a habitat of an endangered species of amphibian. Physicists would develop a computer model to understand the extent of soil contamination and subsurface transport of solvent, chemists would analyze the molecular bonding of the solvent to the specific soil type and biologists would study the impacts upon soil arthropods, plants and ultimately pond dwelling copepods who are the food of the endangered amphibian.

Geosciences includes environmental geology, environmental soil science, volcanic phenomena and evolution of the earth's crust. In some classification systems it can also embrace hydrology including oceanography.

As an example study of soils erosion, calculations would be made of surface runoff by soil scientists. Hydrologists would assist in examining sediment transport in overland flow. Physicists would contribute by assessing the changes in light transmission in the receiving waters. Biologists would analyze subsequent impacts to aquatic flora and fauna from increases in water turbidity.

REGULATIONS DRIVING THE STUDIES

In the U.S. the National Environmental Policy Act {NEPA) of 1969 set forth requirements for analysis of major projects in terms of specific environmental criteria. Numerous state laws have echoed these mandates, applying the principles to local scale actions. The upshot has been an explosion of documentation and study of environmental consequences before the fact of development actions.

One can examine the specifics of environmental science by reading examples of Environmental Impact Statements prepared under NEPA such as: Wastewater treatment expansion options discharging into the San Diego/Tiajuana Estuary, Expansion of the San Francisco International Airport, Development of the Houston, Metro Transportation system, Expansion of the metropolitan Boston MBTA transit system, and Construction of Interstate 66 through Arlington, Virginia.

In England and Wales the Environmental Agency[ ](EA), (EA) formed in 1996 is the leading public body for protecting and improving the environment and enforces the regulations listed on the communities and local government site[ ] (formerly the office of the deputy prime minister) which help drive the study environmental science in the UK. The Agency was set up under the Environment Act 1995 as an independant body and works closely with UK Government to enforce the regulations.

MEANING AND DEFINITION

One can distinguish between "environmental science" and "ecology", since the two fields embody different foci of investigation, although there is overlap. Ecology and environmental science both focus on the interactions of populations of organisms with their environment (including parameters such as meteorology, environmental noise, water quality, air quality, and soil chemistry). A dissimilarity is that ecology is often concerned with issues of biodiversity and the distribution of organisms, while environmental science may address interactions of purely physical parameters that do not involve biological systems. One can consider "ecology" as a basic sub-category to environmental science in analyzing biological systems, both on local and global scales.

ENVIRONMENTAL TECHNOLOGY

Environmental technology or "green technology" is the application of the environmental sciences to conserve the natural environment and resources, and by curbing the negative impacts of human involvement. Sustainable development is the core of environmental technologies. When applying sustainable development as a solution for environmental issues, the solutions need to be socially equitable, economically viable, and environmentally sound. Some environmental technologies that retain sustainable development are; recycling, water purification, sewage treatment, remediation, flue gas treatment, solid waste management, renewable energy, and others.

Related Technologies

Some technologies assist directly with energy conservation, while other technologies are emerging that help the environment by reducing the amount of waste produced by human activities. Energy sources such as solar power create less problems for the environment than traditional sources of energy like coal and petroleum. Scientists continue to search for clean energy alternatives to our current power production methods. Some technologies such as anaerobic digestion produce renewable energy from waste materials. The global reduction of greenhouse gases is dependent on the adoption of energy conservation technologies at industrial level as well as this clean energy generation. The introduction of an energy saving module to common industrial refrigeration or air conditioning can dramatically cut the amount of carbon dioxide (CO_2) and sulphur dioxide (SO_2)that would otherwise be introduced to the atmosphere.

8

Energy Development

Energy Portal

Energy development is the ongoing effort to provide sustainable, accessible energy resources through knowledge, skills, and constructions. When harnessing energy from primary energy sources and converting them into more convenient secondary energy forms, such as electrical energy and cleaner fuels, both emissions (reducing pollution) and quality (more efficient use) are important.

Dependence on External Energy Sources

Technologically advanced societies have become increasingly vulnerable in their dependence on external energy sources for the production of many manufactured goods and the delivery of myriad services. This energy allows people, in general, to live under diverse climatic conditions, in large populations, and often in controlled environments. Level of dependence of human societies on external energy sources differs, as does the climate, comfort, obesity, pollution, output, and greenhouse gas emissions of each society.

Increased levels of human comfort generally induce increased dependence on external energy sources. Conversely, comfort can also be realized with lowered energy consumption by the application of energy efficiency and conservation approaches. Wise energy use therefore embodies the idea of balancing human comfort with reasonable energy consumption levels by researching and implementing effective and sustainable energy harvesting and utilisation measures.

As an example of external energy dependence, U.S. President George Bush has stated that "America is addicted to oil, which is often imported from unstable parts of the world." Another example is the growing reliance on energy supplies to Europe from Russia.

Limitations to Energy Development

Use of any given energy source in human societies encounters limits to quantitative expansion. At the beginning of the 21st century some issues have achieved global dimension. Principal fossil energy sources, such as oil and natural gas are approaching production declines that may occur within the span of a generation. Closely linked to energy development are concerns about the environmental effects of energy use, such as climate changes. Energy development issues are part of the much debated sustainable development problem.

ENERGY SOURCES

Energy sources are substances or processes with concentrations of energy at a high enough potential to be feasibly encouraged to convert to lower energy forms under human control for human benefit. Except for nuclear fuels, tidal energy and geothermal energy, all terrestrial energy sources are solar. And ultimately, both solar energy itself, and geothermal energy, are nuclear in origin.

Fossil Fuels

Fossil fuels, in terms of energy, involve the burning of coal or hydrocarbon fuels, which are the remains of the decomposition of plants and animals. Steam power plant combustion heats water to create steam, which turns a turbine, which, in turn, generates electricity, waste heat, and pollution. There are three main types of fossil fuels: coal, petroleum, and natural gas.

Pros

A refuelling truck containing aviation fuel.

- Because it is based on the simple process of combustion, the burning of fossil fuels can generate large amounts of electricity with a small amount of fuel. Gas-fired power plants are more efficient than coal fired power plants.

- Fossil fuels such as coal are readily available and are currently plentiful. If more energy is required, more coal can be strip-mined. The transport of coal is by rail, especially if the plant is located close to the fuel source.
- Excluding external costs, coal is less expensive than most other sources of energy. Because there are large deposits of coal in the world, the per-unit cost is relatively low.
- The technology already exists for the use of fossil fuels, meaning consumers do not have to spend money switching to other technologies, except for oil and natural gas, as they approach peak production.
- Commonly used fossil fuels in liquid form such as light sweet crude, gasoline, and liquified propane gas are easy to distribute.

Cons

Liquid fossil fuels are highly flammable and their storage can be disastrous when accidents occur.

- The combustion of fossil fuels leads to the release of pollution into the atmosphere. A typical coal plant produces:
  - o 3,700,000,000,000 tons of carbon dioxide (CO_2), the primary human cause of global warming.
  - o 10,000 tons of sulfur dioxide (SO_2), the leading cause of acid rain
  - o 500 tons of small airborne particles, which result in chronic bronchitis, aggravated asthma, and premature death, in addition to haze-obstructed visibility.
  - o 10,200 tons of nitrogen oxide (NOx), leading to formation of ozone (smog) which inflames the lungs, burning through lung tissue making people more susceptible to respiratory illness.

- o 720 tons of carbon monoxide (CO), resulting in headaches and additional stress on people with heart disease.
- o 220 tons of hydrocarbons, volatile organic compounds (VOC), which form ozone.
- o 170 pounds of mercury, where just 1/70th of a teaspoon deposited on a 25-acre lake can make the fish unsafe to eat.
- o 225 pounds of arsenic, which will cause cancer in one out of 100 people who drink water containing 50 parts per billion.
- o 114 pounds of lead, 4 pounds of cadmium, other toxic heavy metals, and trace amounts of uranium.

- Dependence on fossil fuels from volatile regions or countries creates security risks for dependent countries. Oil dependence in particular has lead to monopolization, war, and socio-political instability.
- They are considered non-renewable resources, which will eventually decline in production and become exhausted, with dire consequences to societies that remain highly dependent on them. Fossil fuels are actually slowly forming continuously, but we are using them up at a rate approximately 100,000 times faster than they are formed.
- Extracting fossil fuels is becoming more difficult as we consume the most accessible fuel deposits. Extraction of fossil fuels is becoming more expensive and more dangerous as mines get deeper and oil rigs go further out to sea.
- Extraction of fossil fuels can result in extensive environmental degradation, such as the strip mining and mountaintop removal of coal.
- The drilling and transportation of petroleum can result in accidents that result in the despoilation of hundreds of kilometers of beaches and the death or elimination of many forms of wildlife in the area.

- The storage of these fuels can result in accidents with explosions and poisoning of the atmosphere and groundwater.

HYDROELECTRIC ENERGY

In hydro energy, the gravitational descent of a river is compressed from a long run to a single location with a dam or a flume. This creates a location where concentrated pressure and flow can be used to turn turbines or water wheels, which drive a mechanical mill or an electric generator. An electric generator, when there is excess energy available, can be run backwards as a motor to pump water back up for later use.

Pros

- Hydroelectric power stations can promptly increase to full capacity, unlike other types of power stations. This is because water can be accumulated above the dam and released to coincide with peaks in demand.
- Electricity can be generated constantly, because there are no outside forces, which affect the availability of water. This is in contrast to wind, solar or tidal power, all of which are far less reliable.
- Hydroelectric power produces no waste or pollution.
- Hydropower is a renewable resource; oil, natural gas, and coal reserves may be depleted over time.
- Hydroelectricity secures a country's access to energy supplies.

Cons

- The construction of a dam can have a serious environmental impact on the surrounding areas. The amount and the quality of water downstream can be affected, which affects plant life both aquatic, and land-based. Because a river valley is being flooded, the delicate local habitat of many species are destroyed, while people living nearby may have to relocate their homes.
- Hydroelectricity can only be used in areas where there is a sufficient supply of water.

- Flooding submerges large forests (if they have not been harvested). If such vegetation decayed, it could release methane, a greenhouse gas.

NUCLEAR ENERGY

Nuclear power stations work similarly to fossil fuel power plants, except for the fact that the heat is produced by the reaction of uranium inside a nuclear reactor. The reactor uses uranium rods, the atoms of which are split in the process of fission, releasing a large amount of energy. The process continues as a chain reaction with other nuclei takes place. The heat released heats water to create steam, which spins a turbine, producing electricity.

Pros

- The process of nuclear fission allows for the production of tremendous amounts of energy from a small amount of fuel. The energy content of a pound of uranium or thorium is equivalent to 3.5 million lbs of coal (1,6 million kilos).
- The cost of making nuclear power is about the same as coal, which is considered very inexpensive.
- Nuclear power plants are heavily guarded with the nuclear reactor inside a reinforced containment building, and thus are relatively impervious to terrorist attack or adverse weather conditions.
- Nuclear power does not produce any air pollution or release carbon dioxide and sulfur dioxide into the atmosphere. Therefore, it does not contribute to global warming or acid rain.

Cons

- Waste produced from nuclear fission of uranium is poisonous, and highly radioactive, requiring maintenance and monitoring at the storage sites. Moreover, the long-term disposal of the long-lived nuclear waste causes serious problems, since (unless the spent fuel is reprocessed) it takes from one to three thousand years for the spent fuel to come back to the natural radioactivity of the uranium ore body that was mined to produce it.

- The operation of an uncontained nuclear reactor near human settlements can be catastrophic, as shown by the Chernobyl accident in the Ukraine (former USSR), where large areas of land were affected by nuclear fallout. Members of the public are hesitant about the safety of nuclear power.
- Building a nuclear power plant requires a huge investment, and the costs of safe disassembling (called decommissioning) after it reaches end of usable life have to be included into the budget.
- There can be connections between nuclear power and nuclear weapon proliferation, since both require large-scale uranium enrichment facilities. An international organization (the IAEA) supervises all reactors.
- Nuclear fuels are non-renewable energy sources, with unknown high concentration ore reserves. There is a large amount of trace concentration nuclear material in seawater and most rocks, however extraction from these is not economical.
- Uranium mining activities produce greenhouse gases which contribute to global warming (although not nearly as much as coal mining).

WIND POWER

This type of energy harnesses the power of the wind to propel the blades of wind turbines. These turbines cause the rotation of magnets, which creates electricity. Wind towers are usually built together on wind farms.

Pros

- Wind power produces no water or air pollution that can contaminate the environment, because there are no chemical processes involved in wind power generation.

Hence, there are no waste by-products, such as carbon dioxide.

- Wind generation is a renewable source of energy, which means that we will never run out of it.
- Wind towers can be beneficial for people living permanently, or temporarily, in remote areas. It may be difficult to transport electricity through wires from a power plant to a far-away location and thus, wind towers can be set up at the remote setting.
- Farming and grazing can still take place on land occupied by wind turbines.
- Those utilizing wind power in a grid-tie configuration will have backup power in the event of a grid outage.
- Due to the ability of wind turbines to coexist within agricultural fields, siting costs are frequently low.

Cons

- Wind power is intermittent in many locations, because consistent wind is needed to ensure continuous power generation. When the wind speed decreases, the turbine lingers and less electricity is generated, thus the production at any time in these places is not fully predictable. In some areas, however, winds are highly reliable, or seasonably predictable.
- Commercial generation requires a wind farm over large areas having an effect on scenery.
- Wind farms, depending on the location and type of turbine, can negatively affect bird migration patterns and pose a danger to the birds themselves. Newer, larger wind turbines have slower moving blades which are visible to birds.
- The effect of large scale wind farms on the climate is unknown, just as the effect of buildings, other manmade structures, and agricultural windbreaks have unknown effects on the climate through the extraction of energy from the prevailing wind.

WAVE POWER

Wave power is the extraction of energy from waves in large bodies of water such as oceans and large lakes. Wave power is a form of renewable energy that is on the rise.

Pros

- Potentially highly abundant for countries with large coastlines.
- Potentially minimal effect on the environment
- Wave power is considered to be a renewable resource.

Cons

- No viable large scale method of energy production exists.
- Limited to coastlines.
- Requires further research, development and investment in infrastructure.

BIOMASS

Biomass production involves using garbage or renewable resources, especially vegetation, like maize, to generate electricity. When garbage decomposes the methane produced is captured in pipes and later burned to produce electricity. Vegetation and wood can be burned directly, like fossil fuels, to generate energy, or processed to form alcohols.

Pros

- Biomass production can be used to burn organic waste products resulting from agriculture. This type of recycling encourages the philosophy that nothing on this Earth should be wasted. The result is less demand on the Earth's resources, and a higher carrying capacity for Earth because non-renewable fossil fuels are not consumed.
- Biomass is abundant on Earth and is generally renewable. In theory, we will never run out of organic waste products as fuel, because we are continuously producing them. In addition, biomass is found throughout

the world, a fact that should alleviate energy pressures in third world nations.

- When methods of biomass production other than direct combustion of plant mass, such as fermentation and pyrolysis, are used, there is little effect on the environment. Alcohols and other fuels produced by these alternative methods biomass are clean burning and are feasible replacements to fossil fuels.

Cons

- Direct combustion without emissions filtering generally leads to air pollution similar to that from fossil fuels.
- Producing liquid fuels from biomass is generally less cost effective than from petroleum, since the production of biomass and its subsequent conversion to alcohols is particularly expensive.
- Some researchers claim that, when biomass crops are the product of intensive farming, ethanol fuel production results in a net loss of energy after one accounts for the fuel costs of fertilizer production, farm equipment, and the distillation process.

HYDROGEN FUEL

Unlike the other energy sources in this article, hydrogen fuel must be manufactured with a net loss of energy. When manufactured from natural gas it is, like gasoline, a derivative fuel; when produced using electricity, it is a form chemical energy storage as in storage batteries. In using hydrogen as a fuel, there are two basic alternatives: (1) a fuel cell can convert the chemicals hydrogen and oxygen into water, and in the process, produce electricity, or (2) hydrogen can be burned (less efficiently than in a fuel cell) in an internal combustion engine (e.g. Mazda RX-8 rotary engine).

Pros

- Hydrogen is colourless, odourless and entirely non-polluting, yielding pure water vapor (with minimal NOx) as exhaust when combusted in air. This eliminates the direct production of exhaust gases that lead to smog, and carbon dioxide emissions that enhance the effect of global warming.

- Hydrogen is the lightest chemical element and has the best energy-to-weight ratio of any fuel. Because of this, hydrogen can be economically competitive with gasoline or diesel as a transportation fuel.
- Hydrogen can be produced anywhere; it can be produced domestically from the decomposition of the most abundant chemical on earth: water. Consequently, countries do not have to rely on OPEC countries for fossil fuels. Hydrogen can be produced from domestic sources and the price can be established within the country.
- Electrolysis combined with fuel-cell regeneration is more than 50% efficient; more efficient than pumped hydro and many other forms of mechanical storage.
- Stationary storage with double-walled tanks is stable over long periods of time; hydrogen which outgases from the interior can be pumped back in.

Cons

- Other than some volcanic emanations, hydrogen does not exists in its pure form in the environment, as a gas, because Earth's gravity is not strong enough to hold it at bay at the existing temperature (temperature provides the escape velocity. Helium also isn't retained.) There is concern that a hydrogen economy based on nonhydrocarbon or unreacted hydrogen sources would negatively affect Earth's overall hydrogen budget due to leaks into the atmosphere, and then from the atmosphere into outer space.

  It is impossible to obtain hydrogen gas without expending energy in the process. There are two ways to manufacture hydrogen;

  - o By electrolysis from water-The process of splitting water into oxygen and hydrogen using electrolysis consumes large amounts of energy. It has been calculated that it takes 1.4 joules of electricity to produce 1 joule of hydrogen (Pimentel, 2002).

- o By breaking down hydrocarbons-mainly methane. If oil or gases are used to provide this energy, fossil fuels are consumed, forming pollution and nullifying the value of using a fuel cell. It would be more efficient to use fossil fuel directly.
- o By reacting water with a metal such as Sodium, Potassium, or Boron. Chemical by-products would be sodium oxide, potassium oxide, and boron oxide. Processes exist which could recycle these elements back into their metal form for re-use with additional energy input, further eroding the energy return on energy invested.

- There is currently a lack of infrastructure and distribution network required to support the widespread use of hydrogen as a fuel. It would cost a lot of money and energy to build hydrogen plants and to replace every car and bus with a hydrogen engine and fuel tank.
- Hydrogen is complicated to handle, store, and transport. It requires heavy, cumbersome tanks when stored as a gas, and complex insulating bottles if stored as a cryogenic liquid. If it is needed at a moderate temperature and pressure, a metal hydride absorber may be needed. Transport is also a problem, because hydrogen leaks effortlessly from containers, reducing the efficiency of the fuel. These hassles make hydrogen power very expensive.
- Current efficient fuel cell designs are expensive since they need Platinum as a catalyst. If we were to replace every Internal combustion engine with a Fuel cell then we could potentially use all the Earth's Platinum reserves in two years.

TIDAL POWER

Tidal energy involves building a dam across the opening to a tidal basin, called an estuary. The dam, called a barrage, is composed of turbines, located within tunnels in the dam that rotate when a tide comes in, generating electricity.

Pros

- Tidal power is free once the dam is built. This is because tidal power harnesses the natural power of tides and does not consume fuel. In addition, the maintenance costs associated with running a tidal station are relatively inexpensive.
- Tides are very reliable because it is easy to predict when high and low tides will occur. The tide goes in and out twice a day usually at the predicted times. This makes tidal energy easy to maintain, and positive and negative spikes in energy can be managed.
- Tidal energy is renewable, because nothing is consumed in the rising of tides. Tidal power relies on the gravitational pull of the Moon and Sun, which pull the sea backwards and forwards, generating tides.

Cons

- It provides power for around 10 hours each day, when the tide is moving in or out of the basin.
- The barrage construction can affect the transportation system in water. Boats may not be able to cross the barrage outside of a lock system.
- The erection of a barrage may affect the aquatic ecosystems surrounding it. The environment affected by the dam is very wide, altering areas numerous miles upstream and downstream. For example, many birds rely on low tides to unearth mud flats, which are used as feeding areas.
- Maximum energy production is limited to 2.5 terawatts. This is the total amount of tidal dissipation or the friction measured by the slowing of the lunar orbit.

SOLAR POWER

Solar power involves using solar cells to convert sunlight into electricity, using sunlight hitting solar thermal panels to convert sunlight to heat water or air, using sunlight hitting a parabolic mirror to heat water (producing steam), or using sunlight entering windows for passive solar heating of a building.

Pros

- Solar power is a renewable resource. As long as the Sun exists, its energy will reach Earth.
- Solar power generation releases no water or air pollution, because there is no combustion of fuels.
- In sunny countries, solar power can be used in remote locations, like a wind turbine. This way, isolated places can receive electricity, when there is no way to connect to the power lines from a plant.
- Solar energy can be used very efficiently for heating (solar ovens, solar water and home heaters) and lighting.

Cons

- Solar power is not always completely predictable because it depends on the amount of sunlight that reaches the Earth at any given time. This precludes solar power generation during the night when sunlight does not reach the part of the Earth in which the cell is located and limits solar power generation when cloud cover scatters portions of the electromagnetic spectrum.
- Some forms of solar power are not currently cost competitive. A photovoltaic power station is expensive to build, about 10% efficient, and the energy payback time-the time necessary for producing the same amount of energy than needed for building the power device-for photovoltaic cells is large, of the order of five years. (HomePower Magazine has calculated the payback time to be between 1.8 and 3.3 years, depending primarily on location).

GEOTHERMAL ENERGY

Geothermal energy harnesses the heat energy present underneath the Earth. The hot rocks heat water to produce steam. When holes are drilled in the region, the steam that shoots up is purified and is used to drive turbines, which power electric generators.

Pros

- Geothermal energy does not produce air or water pollution if performed correctly
- Once a geothermal power station is implemented, the energy produced from the station is practically free. A small amount of energy is required in order to run a pump, although this pump can be powered by excess energy generated at the plant.
- Geothermal powers stations are relatively small, and have a lesser impact on the environment than tidal or hydroelectric plants. Because geothermal technology does not rely on large bodies of water, but rather, small, but powerful jets of water, like geysers, large generating stations can be avoided without losing functionality.

Cons

- Geothermal energy is only sufficient as source of power in certain areas of the world. These regions require the presence of hot rocks near the surface to warm the water. The depth of these rocks must be shallow enough that one can drill down to them, and the type of rock also plays a role as it must be easy to drill through.
- Some geothermal sites are prone to running out of steam, when their water is not heated at a high enough temperature to generate steam pressure. This can render the site useless in terms of energy production for decades.
- Drilling holes underground may release hazardous gases and minerals from deep inside the Earth. It can be problematic to dispose of these subsidiary products in a safe manner.

ENERGY TRANSPORTATION

While new sources of energy are only rarely discovered or made possible by new technology, distribution technology continually evolves. The use of fuel cells in cars, for example, is an anticipated delivery technology. This section presents some of the more common delivery technologies that have been important to historic energy development. They all rely in some way on the energy sources listed in the previous section.

- *Fuels:* Shipping is a flexible delivery technology that is used in the whole range of energy development regimes from primitive to highly advanced. Currently, coal, petroleum and their derivatives are delivered by shipping via boat, rail, or road. Petroleum and natural gas may also be delivered via pipeline. Refined hydrocarbon fuels such as gasoline and LPG may also be delivered via aircraft. Natural gas pipelines must maintain a certain minimum pressure to function correctly
- *Electric Grids:* Electricity grids are the networks used to transmit and distribute power from production source to end user, when the two may be hundreds of kilometres away. Sources include electrical generation plants such as a nuclear reactor, coal burning power plant, etc. A combination of sub-stations, transformers, towers, cables, and piping are used to maintain a constant flow of electricity.

Grids may suffer from transient blackouts and brownouts, often due to weather damage. During certain extreme space weather events solar wind can interfere with transmissions.

Grids also have a predefined carrying capacity or load that cannot safely be exceeded. When power requirements exceed what's available, failures are inevitable. To prevent problems, power is then rationed.

Industrialised countries such as Canada, the US, and Australia are among the highest per capita consumers of electricity in the world, which is possible thanks to a widespread electrical distribution network. The US grid is one of the most advanced, although infrastructure maintenance is becoming a problem. The electrical power industry is one of the most heavily subsidized.

Current energy provides a realtime overview of the electricity supply and demand for California, Texas, and the Northeast of the US. African countries with small scale electrical grids have a correspondingly low annual per capita usage of electricity. One of the most powerful power grids in the world supplies power to the state of Queensland, Australia.

ENERGY STORAGE

While most fuels can be stored, electricity in itself cannot. For that reason, many methods of energy storage have been developed, which transform electrical energy into other forms of energy. A method of energy storage may be chosen based on stability, ease of transport, ease of energy release, or ease of converting free energy from the natural form to the stable form.

- Chemical: Some natural forms of energy are found in stable chemical compounds such as fossil fuels. Most systems of chemical energy storage result from biological activity, which store energy in chemical bonds. Man-made forms of chemical energy storage include hydrogen fuel, batteries and explosives such as cordite and dynamite.
- Gravitational: Dams can be used to store energy, by using excess energy to pump water into the reservoir. When electrical energy is required, the process is reversed. The water then turns a turbine, generating electricity. Hydroelectric power is currently an important part of the world's energy supply, generating one-fifth of the world's electricity.

Another example of gravitational energy storage is the counter-weight on elevators.

- Electrical capacitance: Electrical energy may be stored in capacitors. These are often used to produce high intensity releases of energy (such as a camera's flash)
- Mechanical
- Pressure: Energy may also be stored pressurized gases or alternatively in a vacuum. Compressed air, for example, may be used to operate vehicles and power tools. Large scale compressed air energy storage facilities are used to smooth out demands on electricity generation

by providing energy during peak hours and storing energy during off-peak hours. Such systems save on expensive generating capacity since it only needs to meet average consumption rather than peak consumption.

- Flywheels and springs: Energy can also be stored in mechanical systems such as springs or flywheels. Flywheel energy storage is currently being used for uninterruptible power supplies.

FUTURE ENERGY DEVELOPMENT

Extrapolations from current knowledge to future energy development offer a choice of energy futures. Some predictions parallel the Malthusian catastrophe hypothesis. Numerous are complex models based scenarios as pioneered by Limits to Growth. Modelling approaches offer ways to analyse diverse strategies, and hopefully find a road to rapid and sustainable development of humanity. Short term energy crises are also a concern of energy development.

Existing technologies for new energy sources, such as new renewable energy technologies, nuclear fission and fusion are promising, but need sustained research and development, including consideration of possible harmful side effects. Artificial Photosynthesis is another energy technology being researched and developed.

WORLD ENERGY RESOURCES AND CONSUMPTION

The world has a number of energy resources which provide us with the ability to perform work (a narrowly-defined measure of the ability to effect physical change). Energy resources range from fossil fuels and nuclear fuels to renewable energy such as wind, hydro-and solar energy. When this energy is drawn upon to do work, and is converted to a less useful form, energy consumption is said to have occured.

World Energy Consumption

In 2004 the worldwide energy consumption of the human race was estimated as 15 TW (TW=1012 Watts) by the United

States Energy Information Administration. This is equivalent to 0.5 ZJ (ZJ =1021J) per year. There is at least 10% uncertainty in this number because the underlying data is continuously changing and not all of the world's economies track their energy consumption with the same rigor. Consumption by fuel type Eighty-seven percent of the world's energy is supplied by fossil fuel. The 15 TW total energy consumption of 2004 was divided as follows:

	TW=10^{12} Watt	ZJ=10^{21}J per year
Oil	5.6	0.18
Gas	3.5	0.11
Coal	3.8	0.12
Hydroelectric	0.9	0.03
Nuclear	0.9	0.03
Geothermal, Wind, Solar, Wood	0.2	0.006

Ever since the advent of the industrial revolution, the worldwide energy consumption has been growing steadily. In 1890 the consumption of fossil fuels roughly equaled the amount of bio mass fuel burned by households and industry. In 1900 global energy consumption equaled 0.7 TW. The twentieth century saw a rapid twenty fold increase in fossil fuels. Between 1980 and 2004 the worldwide annual growth rate has been 2%. The most significant growth of energy consumption is currently taking place in China which has been growing at 5.5% over the last 25 years. Its 1.3 billion people are currently consuming energy at a rate of 2 KW per person. The growth in energy consumption over the last quarter century. Over the last twenty years Fossil fuels have continued to grow and increased their share of the energy supply, coal has been growing the fastest. Nuclear and hydro electric energy have stagnated due to environmental concerns. With the exception of France no major western country has ordered a new nuclear power plant since 1980.

In 2004 renewable energy supplied around 7% of the world's energy consumption. Most (87%) of the renewable energy was hydroelectric power. Solar and wind power provided 4 and 65 GW (GW=109 Watt) respectively. Consumption by country energy consumption broadly tracks with gross national product, although there is a significant difference between the consumption levels of the US with 11.4 KW per person, and Japan and Germany's with 6 KW per person. In developing countries such as India the per person energy use is closer to 0.5 KW. The per capita energy consumption against the GDP per person for all countries with more than 20 million inhabitants, this covers 95% of the world population. Canada has the highest energy consumption per capita whereas the lowest energy consumption takes place in the third world.

Energy Consumption by Sector

Transportation, industry, residential and commercial (offices & shops) each consume approximately equal shares of the total 15 TW.A third of the world's energy is used to produce electricity. In 2005, global electricity consumption equaled 2 TW. The energy used to generate 2 TW of electricity is approximately 5 TW, as the efficiency of a typical existing power plant is around 38%. The new generation of gas fired plants reaches a substantially higher efficiency of 55%. Coal is the most popular fuel for the world's electricity plants. World energy resources Fossil fuel resources. Despite several voices predicting the imminent decline of fossil fuels, there are still significant reserves of all the traditional energy components. Remaining reserves of conventional fossil fuels are estimated as:

Coal	290 ZJ
Oil	57 ZJ
Gas	30 ZJ

Again significant uncertainty exists for these numbers. This uncertainty is not surprising, since the estimation for the remaining fossil fuels on the planets depends on a detailed understanding of the earth crust. This understanding is still less than perfect. While modern drilling technology makes it possible to drill wells in up to 3000 meters of water to verify the exact composition of the geology, half of the ocean has depth exceeding 3000 meters, leaving about a third of the planet beyond the reach of detailed analysis. The make up of the estimated 57 ZJ of oil on earth. The estimations for the world's oil reserves vary from low of 8 ZJ, consisting of currently proven and recoverable reserves, to a maximum of 110 ZJ consisting of available, but not necessary recoverable reserves, and including optimistic estimates for unconventional sources such as tar sands and oil shale. Both estimates provide oil for the foreseeable future at current oil consumption rate of 0.18 ZJ per year. There is a broad consensus among scientists that we are not close to running out of fossil fuels.

Coal is especially abundant and by itself can sustain the current energy consumption of the entire planet for the next 600 years. Renewable energy resources political and/or environmental considerations might move the world's energy consumption away from fossil fuels. For example, the European Commission has proposed in its Renewable Energy Roadmap, a binding target of increasing the level of renewable energy in the EU's overall mix from less than 7% today to 20% by 2020.

In 2005 the Swedish government announced their intention to become the first country to break their dependence on oil and other 'fossil raw materials' by 2020. Renewable energy sources are even larger than the traditional fossil fuels and in theory can easily supply our energy needs. The sunlight interacts with the planet. There is 89,000 TW of solar energy that falls on the planet's surface. In other words, we need to capture less than 0.02% of the available solar energy to meet our current energy needs. The available wind energy estimates range from 300 to 370 TW.

Using the lower estimate, just 5% of the available wind energy would supply the current worldwide energy needs. Most of this wind energy is available over open ocean. Not only does the ocean cover 71% of the planet, wind also tends to blow stronger over open water because there are less obstructions. The above percentages for solar and wind energy ignore the formidable challenges of energy distribution and storage that need to be solved to overcome the intermittent and seasonal variations of these energy sources. The numbers do however illustrate that it is technically possible to move away from fossil fuels should society decide to make the necessary investment.

The required investment to change energy sources is enormous. It is estimated that the cost to replace the global infrastructure for liquid transportation fuels alone is $3-5 trillion.

Alternative Energy Paths

Whether a significant investment in renewable energy is wise or not, is currently the subject of much debate. Trials are taking place in several countries; Japan and Germany have started to make some investments in solar energy, making them the largest consumers of photovoltaic cells in the world despite their unfavorable geographic location.

Denmark and Germany have installed 3 GW and 17 GW of wind power respectively. In 2005 wind generated 18.5% of all the electricity in Denmark. Brazil has made a significant investment in ethanol fuel which supplies a significant part of the transportation fuel in that country. Starting in 1965, France made large investments in nuclear power and to this date three quarters of their electricity comes from their nuclear pressure water reactor plants. In the 21st century some of these different energy paths might become more mainstream and start replacing the ubiquitous fossil fuels.

EARTH'S ENERGY BUDGET

A schematic representation of the energy exchanges between the Earth's surface, the Earth's atmosphere, and outer space. The recycling of energy between the Earth's surface and the atmosphere is characteristic of the greenhouse effect. Note that the total energy entering each level is equal to the energy leaving that level as should be expected for a system in balance.

The Earth can be considered as a physical system with an energy budget that includes all gains of incoming energy and all losses of outgoing energy. The planet is approximately in equilibrium, so the sum of the gains should be approximately equal to the sum of the losses.

Note: Although the term "energy budget" is widely used, the flow of energy in and out of the Earth is actually measured in units of power (watts), not units of energy (joules). Therefore, "power budget" would be a more accurate term.

The Power Budget

The total flux of power entering the Earth's atmosphere is estimated at 174 peta watts. This consists of:

- solar radiation (99.978%, or nearly 174 petawatts; or about 340 W m-2)
  - o This is equal to the product of the solar constant, about 1366 watts per square metre, and the area of the Earth's disc as seen from the Sun, about 1.28 × 1014 square metres, averaged over the Earth's surface, which is four times larger. The solar flux averaged over just the sunlit half of the earth's surface is about 680 W m-2
- geothermal energy (0.013%, or about 23 terawatts; or about 0.045 W m-2)
  - o This is produced by stored heat and heat produced by radioactive decay leaking out of the Earth's interior.
- tidal energy (0.002%, or about 3 terawatts; or about 0.0059 W m-2)
  - o This is produced by the interaction of the Earth's mass with the gravitational fields of other bodies such as the Moon and Sun.
- waste heat from fossil fuel consumption (about 0.007%, or about 13 terawatts; or about 0.025 W m-2).

Note that the solar constant varies (by approximately 0.1% over a solar cycle); and is not known absolutely to within better than about one watt per square metre. Hence the geothermal and tidal contributions are less than the uncertainty in the solar power.

Outgoing Power

An average albedo (reflectivity) of the Earth is about 0.3, which means that 30% of the incident solar energy is reflected back into space, while 70% is absorbed by the Earth and reradiated as infrared. The planet's albedo varies from month to month, but 0.3 is the average figure. It also varies very strongly spatially: ice sheets have a high albedo, oceans low.

The contributions from geothermal and tidal power sources are so small that they are omitted from the following calculations.

The 30% reflected energy consists of:

- 6% reflected from the atmosphere
- 20% reflected from clouds
- 4% reflected from the ground (including land, water and ice)

All of the 70% absorbed energy is eventually reradiated:

- 64% by the clouds and atmosphere
- 6% by the ground

The same 70% of absorbed energy can be split this way:

- 51% absorbed by land and water, then emerging in the following ways:
  - o 23% transferred back into the atmosphere as latent heat by the evaporation of water
  - o 7% transferred back into the atmosphere by heated rising air
  - o 6% radiated directly into space
  - o 15% transferred into the atmosphere by radiation, then reradiated into space, and
- 19% absorbed by the atmosphere, including:
  - o 16% reradiated back into space
  - o 3% transferred to clouds, from where it is radiated back into space

Anthropogenic Modification

Emission of greenhouse gases, and other factors (such as land-use changes), modify the energy budget slightly but significantly. The IPCC provides an estimate of these forcing, insofar as they are known. The largest and best known are from the well-mixed greenhouse gases (CO_2, CH_4, halocarbons, etc.), totalling an increase in forcing of 2.4 W·m-2 relative to 1750. These are less than 1% of the solar input, but contributes to the observed increase in atmospheric and oceanic temperature.

Electricity Generation

Electricity generation is the first process in the delivery of electricity to consumers. The other processes are electric power transmission and electricity distribution.

The importance of dependable electricity generation, transmission and distribution was revealed when it became apparent that electricity was useful for providing heat, light and power for human needs. Centralized power generation became possible when it was recognized that alternating current electric power lines can transport electricity at low costs across great distances by taking advantage of the ability to transform the voltage using power transformers.

Electricity has been generated for the purpose of powering human technologies for at least 120 years from various sources of energy. The first power plants were run on wood, while today we rely mainly on petroleum, natural gas, coal, hydroelectric and nuclear power and a small amount from hydrogen, solar energy, tidal harnesses, wind generators, and geothermal sources.

Electricity Demand

The demand for electricity can be met in two different ways. The primary method thus far has been for public or private utilities to construct large scale centralized projects to generate and transmit the electricity required to fuel economies. Many of these projects have caused unpleasant environmental effects such as air or radiation pollution and the flooding of large areas of land.

Distributed generation creates power on a smaller scale at locations throughout the electricity network. Often these sites generate electricity as a by-product of other industrial processes such as using gas from landfills to drive turbines.

METHODS OF GENERATING ELECTRICITY

Turbines

Rotating turbines attached to electrical generators produce most commercially available electricity. Turbines are driven by a fluid which acts as an intermediate energy carrier. The fluids typically used are:

- *Steam:* Water is boiled by nuclear fission or the burning of fossil fuels (coal, natural gas, or petroleum). Some newer plants use the sun as the heat source: solar parabolic troughs and solar power towers concentrate sunlight to heat a heat transfer fluid, which is then used to produce steam.
- *Water:* Turbine blades are acted upon by flowing water, produced by hydroelectric dams or tidal forces,
- *Wind:* Most wind turbines generate electricity from naturally occurring wind. Solar updraft towers use wind that is artificially produced inside the chimney by heating it with sunlight.
- *Hot Gases:* Turbines are driven directly by gases produced by the combustion of natural gas or oil.

Combined cycle gas turbine plants are driven by both steam and gas. They generate power by burning natural gas in a gas turbine and use residual heat to generate additional electricity from steam. These plants offer efficiencies of up to 60%.

Reciprocating Engines

Small electricity generators are often powered by reciprocating engines burning diesel, biogas or natural gas. Diesel engines are often used for back up generation, usually at low voltages. Biogas is often combusted where it is produced, such as a landfill or wastewater treatment plant, with a reciprocating engine or a microturbine, which is a small gas turbine.

Photovoltaic Panels

Unlike the solar heat concentrators mentioned above, photovoltaic panels convert sunlight directly to electricity. Although sunlight is free and abundant, solar panels are expensive to produce and have only a 10-20% conversion efficiency. Until recently, photovoltaics were most commonly used in remote sites where there is no access to a commercial power grid, or as a supplemental electricity source for individual homes and businesses. Recent advances in manufacturing efficiency and photovoltaic technology, combined with subsidies driven by environmental concerns, have dramatically accelerated the deployment of solar panels. Installed solar capacity is growing by 30% per year in several regions including Germany, Japan, California and New Jersey.

Nuclear Power

A nuclear power station. Condensed water vapor rises from the hyperboloid shaped cooling towers. The nuclear reactors are inside the cylindrical containment buildings.

Nuclear power is the controlled use of nuclear reactions to release energy for work including propulsion, heat, and the generation of electricity. Human use of nuclear power to do significant useful work is currently limited to nuclear fission and radioactive decay. Nuclear energy is produced when a fissile material, such as uranium-235 (235U), is concentrated such that nuclear fission takes place in a controlled chain reaction and creates heat-which is used to boil water, produce steam, and drive a steam turbine. The turbine can be used for mechanical work and also to generate electricity. Nuclear power is used to power most military submarines and aircraft carriers and provides 7% of the world's energy and 15.7% of the world's electricity.

The United States produces the most nuclear energy, with nuclear power providing 20% of the electricity it consumes, while France produces the highest percentage of its electrical energy from nuclear reactors-80% as of 2006. Nuclear energy policy differs between countries.

Nuclear energy uses an abundant, widely distributed fuel, and mitigates the greenhouse effect if used to replace fossil-fuel-derived electricity. International research is ongoing into various safety improvements, the use of nuclear fusion and additional uses such as the generation of hydrogen (in support of hydrogen economy schemes), for desalinating sea water, and for use in district heating systems. Construction of nuclear power plants in the U.S. declined following the 1979 Three Mile Island accident and the 1986 disaster at Chernobyl. Lately, there has been renewed interest in nuclear energy from national governments due to economic and environmental concerns. Other reasons for interest include the public, some notable environmentalists due to increased oil prices, new passively safe designs of plants, and the low emission rate of greenhouse gas which some governments need to meet the standards of the Kyoto Protocol. A few reactors are under construction, and several new types of reactors are planned.

The use of nuclear power is controversial because of the problem of storing radioactive waste for indefinite periods, the potential for possibly severe radioactive contamination by accident or sabotage, and the possibility that its use in some countries could lead to the proliferation of nuclear weapons. Proponents believe that these risks are small and can be further reduced by the technology in the new reactors. They further claim that the safety record is already good when compared to other fossil-fuel plants, that it releases much less radioactive waste than coal power, and that nuclear power is a sustainable energy source. Critics, including most major environmental groups, believe nuclear power is an uneconomic, unsound and potentially dangerous energy source, especially compared to renewable energy, and dispute whether the costs and risks can be reduced through new technology. There is concern in some countries over North Korea and Iran operating research reactors and fuel enrichment plants, since those countries refuse adequate IAEA oversight and are believed to be trying to develop nuclear weapons. North Korea admits that it is developing nuclear weapons, while the Iranian government vehemently denies the claims against Iran.

Origins

The first successful experiment with nuclear fission was conducted in 1938 in Berlin by the German physicists Otto Hahn, Lise Meitner and Fritz Strassmann.

During the Second World War, a number of nations embarked on crash programs to develop nuclear energy, focusing first on the development of nuclear reactors. The first self-sustaining nuclear chain reaction was obtained at the University of Chicago by Enrico Fermi on December 2, 1942, and reactors based on his research were used to produce the plutonium necessary for the "Fat Man" weapon dropped on Nagasaki, Japan. Several nations began their own construction of nuclear reactors at this point, primarily for weapons use, though research was also being conducted into their use for civilian electricity generation.

Electricity was generated for the first time by a nuclear reactor on December 20, 1951 at the EBR-I experimental fast breeder station near Arco, Idaho, which initially produced about 100 kW.

In 1952 a report by the Paley Commission (The President's Materials Policy Commission) for President Harry Truman made a "relatively pessimistic" assessment of nuclear power, and called for "aggressive research in the whole field of solar energy".

A December 1953 speech by President Dwight Eisenhower, "Atoms for Peace", set the U.S. on a course of strong government support for the international use of nuclear power.

Early Years

The Beaver Valley Nuclear Generating Station in Shippingport, Pennsylvania was the first commercial reactor in the USA and was opened in 1957.

On June 27, 1954, the world's first nuclear power plant to generate electricity for a power grid started operations at Obninsk, USSR. The reactor was graphite moderated, water cooled and had a capacity of 5 megawatts (MW). The world's first commercial nuclear power station, Calder Hall in Sellafield, England was opened in 1956, a gas-cooled Magnox reactor with an initial capacity of 50 MW (later 200 MW). The Shippingport Reactor (Pennsylvania, 1957), a pressurized water reactor, was the first commercial nuclear generator to become operational in the United States.

In 1954, the chairman of the United States Atomic Energy Commission (forerunner of the U.S. Nuclear Regulatory Commission) talked about electricity being "too cheap to meter" in the future, often misreported as a concrete statement about nuclear power, and foresaw 1000 nuclear plants on line in the USA by the year 2000.

In 1955 the United Nations' "First Geneva Conference", then the world's largest gathering of scientists and engineers, met to explore the technology. In 1957 EURATOM was launched alongside the European-Economic Community (the latter is now the European Union). The same year also saw the launch of the International Atomic Energy Agency (IAEA).

Development

Installed nuclear capacity initially rose relatively quickly, rising from less than 1 gigawatt (GW) in 1960 to 100 GW in the late 1970s, and 300 GW in the late 1980s. Since the late 1980s capacity has risen much more slowly, reaching 366 GW in 2005, primarily due to Chinese expansion of nuclear power. Between around 1970 and 1990, more than 50 GW of capacity was under construction (peaking at over 150 GW in the late 70s and early 80s)-in 2005, around 25 GW of new capacity was planned. More than two-thirds of all nuclear plants ordered after January 1970 were eventually cancelled.

During the 1970s and 1980s rising economic costs (related to vastly extended construction times largely due to regulatory delays) and falling fossil fuel prices made nuclear power plants then under construction less attractive. In the 1980s (U.S.) and 1990s (Europe), flat load growth and electricity liberalization also made the addition of large new baseload capacity unnecessary.

A general movement against nuclear power arose during the last third of the 20th century, based on the fear of a possible nuclear accident and on fears of latent radiation, and on the opposition to nuclear waste production, transport and final storage.

Perceived risks on the citizens' health and safety, the 1979 accident at Three Mile Island and the 1986 Chernobyl accident played a key part in stopping new plant construction in many countries. Austria (1978), Sweden (1980) and Italy (1987) voted in referendums to oppose or phase out nuclear power, while opposition in Ireland prevented a nuclear programme there. However, the Brookings Institution suggests that new nuclear units have not been ordered primarily for economic reasons rather than fears of accidents.

Financing for new reactors dried up when Wall Street's enthusiasm ended. Disillusionment was complete when new research discredited the claim (previously accepted as fact even by opponents) that nuclear power was still, despite all its problems, the most cost-effective source of electricity. Industry figures had omitted the factor of downtime. During the 1980s and early 1990s, the newest and biggest U.S. plants were actually producing only half the energy they were supposed to, due to shutdowns for refueling, routine maintenance, retrofitting, and frequent minor mishaps. Since that time, the capacity factor of existing nuclear power plants has increased dramatically, and has been near 90% in the current decade.

As of 2006, the stated desire to use nuclear power for electricity generation has been suspected of being a cover for nuclear proliferation in the countries of Iran and North Korea.

Reactor Types

Current Technology: There are two types of nuclear power in current use:

1. The nuclear fission reactor produces heat through a controlled nuclear chain reaction in a critical mass of fissile material.

All current nuclear power plants are critical fission reactors, which are the focus of this article. The output of fission reactors is controllable. There are several subtypes of critical fission reactors, which can be classified as Generation I, Generation II and Generation III. All reactors will be compared to the Pressurized Water Reactor (PWR), as that is the standard modern reactor design. The difference between fast-spectrum and thermal-spectrum reactors will be covered later. In general, fast-spectrum reactors will produce less waste, and the waste they do produce will have a vastly shorter halflife, but they are more difficult to build, and more expensive to operate. Fast reactors can also be breeders, whereas thermal reactors generally cannot.

Pressurized Water Reactors (PWR)

These are reactors cooled and moderated by high pressure liquid (even at extreme temperatures) water. They are the majority of current reactors, and are generally considered the safest and most reliable technology currently in large scale deployment, although Three Mile Island is a reactor of this type. This is a thermal neutron reactor design.

Boiling Water Reactors (BWR)

These are reactors cooled and moderated by water, under slightly lower pressure. The water is allowed to boil in the reactor. The thermal efficiency of these reactors can be higher, and they can be simpler, and even potentially more stable and safe. Unfortunately, the boiling water puts more stress on many of the components, and increases the risk that radioactive water may escape in an accident. These reactors make up a substantial percentage of modern reactors. This is a thermal neutron reactor design.

Pressurized Heavy Water Reactor (PHWR)

A Canadian design, (known as CANDU) these reactors are heavy-water-cooled and-moderated Pressurized-Water reactors. Instead of using a single large containment vessel as in a PWR, the fuel is contained in hundreds of pressure tubes. These reactors are fuelled with natural uranium and are thermal neutron reactor designs.

PHWRs can be refueled while at full power, which makes them very efficient in their use of uranium (it allows for precise flux control in the core). Most PHWRs exist within Canada, but units have been sold to Argentina, China, India (pre-NPT), Pakistan (pre-NPT), Romania, and South Korea. India also operates a number of PHWR's, often termed 'CANDU-derivatives', built after the 1974 Smiling Buddha nuclear weapon test.

Reaktor Bolshoy Moshchnosti Kanalniy (RBMK)

A Soviet Union design, built to produce plutonium as well as power, the dangerous and unstable RBMKs are water cooled with a graphite moderator. RBMKs are in some respects similar to CANDU in that they are refuelable On-Load and employ a pressure tube design instead of a PWR-style pressure vessel. However, unlike CANDU they are very unstable and too large to have containment buildings. Because of this RBMK reactors are generally considered one of the most dangerous reactor designs in use. Chernobyl was an RBMK.

Gas Cooled Reactor (GCR) and Advanced Gas Cooled Reactor (AGCR)

These are generally graphite moderated and CO2 cooled. They have a high thermal efficiency compared with PWRs and an excellent safety record. There are a number of operating reactors of this design, mostly in the United Kingdom. Older designs (i.e. Magnox stations) are either shut down or will be in the near future. However, the AGCRs have an anticipated life of a further 10 to 20 years. This is a thermal neutron reactor design.

Super Critical Water-cooled Reactor (SCWR)

This is a theoretical reactor design that is part of the Gen-IV reactor project. It combines higher efficiency than a GCR with the safety of a PWR, though it is perhaps more technically challenging than either. The water is pressurized and heated past its critical point, until there is no difference between the liquid and gas states. An SCWR is similar to a BWR, except there is no boiling (as the water is critical), and the thermal efficiency is higher as the water behaves more like a classical gas. This is an epithermal neutron reactor design.

Liquid Metal Fast Breeder Reactor (LMFBR)

This is a reactor design that is cooled by liquid metal, totally unmoderated, and produces more fuel than it consumes. These reactors can function much like a PWR in terms of efficiency, and do not require much high pressure containment, as the liquid metal does not need to be kept at high pressure, even at very high temperatures.

Superphénix in France was a reactor of this type, as was Fermi-I in the United States. The Monju reactor in Japan suffered a sodium leak in 1995 and is approved for restart in 2008. All three use/used liquid sodium. These reactors are fast neutron, not thermal neutron designs. These reactors come in two types:

Lead Cooled

Using lead as the liquid metal provides excellent radiation shielding, and allows for operation at very high temperatures. Also, lead is (mostly) transparent to neutrons, so fewer neutrons are lost in the coolant, and the coolant does not become radioactive. Unlike sodium, lead is mostly inert, so there is less risk of explosion or accident, but such large quantities of lead may be problematic from toxicology and disposal points of view. Often a reactor of this type would use a lead-bismuth eutectic mixture. In this case, the bismuth would present some minor radiation problems, as it is not quite as transparent to neutrons, and can be transmuted to a radioactive isotope more readily than lead.

Sodium Cooled

Most LMFBRs are of this type. The sodium is relatively easy to obtain and work with, and it also manages to actually remove corrosion on the various reactor parts immersed in it. However, sodium explodes violently when exposed to water, so care must be taken, but such explosions wouldn't be vastly more violent than (for example) a leak of superheated fluid from a SCWR or PWR.

The radioisotope thermoelectric generator produces heat through passive radioactive decay.

Some radioisotope thermoelectric generators have been created to power space probes (for example, the Cassini probe), some lighthouses in the former Soviet Union, and some pacemakers. The heat output of these generators diminishes with time; the heat is converted to electricity utilising the thermoelectric effect.

Working of Nuclear Power Plants

The key components common to most types of nuclear power plants are:

- Nuclear fuel
- Neutron moderator
- Coolant

- Control rods
- Pressure vessel
- Emergency core cooling systems
- Reactor protective system
- Steam generators (not in BWRs)
- Containment building
- Boiler feedwater pump
- Turbine
- Electrical generator
- Condenser

Conventional thermal power plants all have a heat source. Examples are gas, coal, or oil. For a nuclear power plant, this heat is provided by nuclear fission inside the nuclear reactor. When a relatively large fissile atomic nucleus (usually uranium-235 or plutonium-239) is struck by a neutron it forms two or more smaller nuclei as fission products, releasing energy and neutrons in a process called nuclear fission.

The neutrons then trigger further fission. And so on. When this nuclear chain reaction is controlled, the energy released can be used to heat water, produce steam and drive a turbine that generates electricity. It should be noted that a nuclear explosive involves an uncontrolled chain reaction, and the rate of fission in a reactor is not capable of reaching sufficient levels to trigger a nuclear explosion because commercial reactor grade nuclear fuel is not enriched to a high enough level.

The chain reaction is controlled through the use of materials that absorb and moderate neutrons. In uranium-fueled reactors, neutrons must be moderated (slowed down) because slow neutrons are more likely to cause fission when colliding with a uranium-235 nucleus. Light water reactors use ordinary water to moderate and cool the reactors. When at operating temperatures if the temperature of the water increases, its density drops, and fewer neutrons passing through it are slowed enough to trigger further reactions. That negative feedback stabilizes the reaction rate.

Experimental Technologies

A number of other designs for nuclear power generation, the Generation IV reactors, are the subject of active research and may be used for practical power generation in the future. A number of the advanced nuclear reactor designs could also make critical fission reactors much cleaner, much safer and/or much less of a risk to the proliferation of nuclear weapons.

- *Integral Fast Reactor:* The link at the end of this paragraph references an interview with Dr. Charles Till, former director of Argonne National Laboratory West in Idaho and outlines the Integral Fast Reactor and its advantages over current reactor design, especially in the areas of safety, efficient nuclear fuel usage and reduced waste. The IFR was built, tested and evaluated during the 1980s and then retired under the Clinton administration in the 1990s due to nuclear non-proliferation policies of the administration. Recycling spent fuel is the core of its design and it therefore produces a fraction of the waste of current reactors.
- *Pebble Bed Reactor:* This reactor type is designed so high temperatures reduce power output by doppler broadening of the fuel's neutron cross-section. It uses ceramic fuels so its safe operating temperatures exceed the power-reduction temperature range. Most designs are cooled by inert helium, which cannot have steam explosions, and which does not easily absorb neutrons and become radioactive, or dissolve contaminants that can become radioactive. Typical designs have more layers (up to 7) of passive containment than light water reactors (usually 3). A unique feature that might aid safety is that the fuel-balls actually form the core's mechanism, and are replaced one-by-one as they age. The design of the fuel makes fuel reprocessing expensive.
- SSTAR, Small, Sealed, Transportable, Autonomous Reactor is being primarily researched and developed in the US, intended as a fast breeder reactor that is tamper resistant and passively safe.

- Subcritical reactors are designed to be safer and more stable, but pose a number of engineering and economic difficulties.
- Controlled nuclear fusion could in principle be used in fusion power plants to produce safer, cleaner power, but significant scientific and technical obstacles remain. Several fusion reactors have been built, but as yet none has 'produced' more thermal energy than electrical energy consumed. Despite research having started in the 1950s, no commercial fusion reactor is expected before 2050. The ITER project is currently leading the effort to commercialize fusion power.
- Thorium based reactors. It is possible to convert Thorium-232 into U-233 in reactors specially designed for the purpose. In this way, Thorium, which is more plentiful than uranium, can be used to breed U-233 nuclear fuel. U-233 is also believed to have favourable nuclear properties as compared to traditionally used U-235, including better neutron economy and lower production of long lived transuranic waste.
- *Advanced Heavy Water Reactor:* A proposed heavy water moderated nuclear power reactor that will be the next generation design of the PHWR type. Under development in the Bhabha Atomic Research Centre (BARC).
- KAMINI-A unique reactor using Uranium-233 isotope for fuel. Built by BARC and IGCAR Uses thorium.
- India is also building a bigger scale FBTR or fast breeder thorium reactor to harness the power with the use of thorium.

Life Cycle

The Nuclear Fuel Cycle begins when uranium is mined, enriched, and manufactured into nuclear fuel, (1) which is delivered to a nuclear power plant. After usage in the power plant, the spent fuel is delivered to a reprocessing plant (2) or to a final repository (3) for geological disposition. In reprocessing 95% of spent fuel can be recycled to be returned to usage in a power plant (4).

Nuclear Fuel

A nuclear reactor is only part of the life-cycle for nuclear power. The process starts with mining. Generally, uranium mines are either open-pit strip mines, or in-situ leach mines. In either case, the uranium ore is extracted, usually converted into a stable and compact form such as yellowcake, and then transported to a processing facility. Here, the yellowcake is converted to uranium hexafluoride, which is then enriched using various techniques. At this point, the enriched uranium, containing more than the natural 0.7% U-235, is used to make rods of the proper composition and geometry for the particular reactor that the fuel is destined for. The fuel rods will spend about 3 years inside the reactor, generally until about 3% of their uranium has been fissioned, then they will be moved to a spent fuel pool where the short lived isotopes generated by fission can decay away. After about 5 years in a cooling pond, the spent fuel is radioactively cool enough to handle, and it can be moved to dry storage casks or reprocessed.

Fuel Resources

Uranium is a common element, occurring almost everywhere on land and in the oceans. It is about as common as tin, and 500 times more common than gold. Most types of rocks and soils contain uranium, although often in low concentrations. At present, economically viable deposits are regarded as being those with concentrations of at least 0.1% uranium. At this cost level, available reserves would last for 50 years at the present rate of use. Doubling the price of uranium, which would have only little effect on the overall cost of nuclear power, would increase reserves to hundreds of years. To put this in perspective; a doubling in the cost of natural uranium would increase the total cost of nuclear power by 5%. On the other hand, if the price of natural gas was doubled, the cost of gas-fired power would increase by about 60%. Doubling the price of coal would increase the cost of power production in a large coal-fired power station by about 30%.

Uranium enrichment produces many tons of depleted uranium (DU) which consists of U-238 with most of the easily fissile U-235 isotope removed. U-238 is a tough metal with several commercial uses-for example, aircraft production, radiation shielding, and making bullets and armor-as it has a higher density than lead. There are concerns that U-238 may lead to health problems in groups exposed to this material excessively, like tank crews and civilians living in areas where large quantities of DU ammunition have been used.

Current light water reactors make relatively inefficient use of nuclear fuel, leading to energy waste. More efficient reactor designs or nuclear reprocessing would reduce the amount of waste material generated and allow better use of the available resources.

As opposed to current light water reactors which use uranium-235 (0.7% of all natural uranium), fast breeder reactors use uranium-238 (99.3% of all natural uranium). It has been estimated that there is up to five-billion years' worth of uranium-238 for use in these power plants. Breeder technology has been used in several reactors. Currently (December 2005), the only breeder reactor producing power is BN-600 in Beloyarsk, Russia. (The electricity output of BN-600 is 600 MW-Russia has planned to build another unit, BN-800, at Beloyarsk nuclear power plant.) Also, Japan's Monju reactor is planned for restart (having been shut down since 1995), and both China and India intend to build breeder reactors.

Another alternative would be to use uranium-233 bred from thorium as fission fuel-the thorium fuel cycle. Thorium is three times more abundant in the Earth's crust than uranium, and (theoretically) all of it can be used for breeding, making the potential thorium resource orders of magnitude larger than the uranium fuel cycle operated without breeding. Unlike the breeding of U-238 into plutonium, fast breeder reactors are not necessary-it can be performed satisfactorily in more conventional plants.

Proposed fusion reactors assume the use of deuterium, an isotope of hydrogen, as fuel and in most current designs also lithium. Assuming a fusion energy output equal to the current global output and that this does not increase in the future, then the known current lithium reserves would last 3,000 years, lithium from sea water would last 60 million years, and a more complicated fusion process using only deuterium from sea water would have fuel for 150 billion years. For comparison, the Sun has an estimated remaining life of 5 billion years.

Solid Waste

The predominant waste stream from nuclear power plants is spent fuel. A large nuclear reactor produces 3 cubic metres (25-30 tonnes) of spent fuel each year. It is primarily composed of unconverted uranium as well as significant quantities of transuranic actinides (plutonium and curium, mostly). In addition, about 3% of it is made of fission products. The actinides (uranium, plutonium, and curium) are responsible for the bulk of the long term radioactivity, whereas the fission products are responsible for the bulk of the short term radioactivity.

Spent fuel is highly radioactive and needs to be handled with great care and forethought. Fresh from the reactor, it is so radioactive that less than a minute's exposure to it will cause death. However, spent nuclear fuel becomes less radioactive over time. After 40 years, the radiation flux is 99.9% lower than it was the moment the spent fuel was removed, although still dangerously radioactive.

The safe storage and disposal of nuclear waste is a significant challenge. Because of potential harm from radiation, spent nuclear fuel must be stored in shielded basins of water (spent fuel pools), and usually subsequently in dry storage vaults or dry cask storage until its radioactivity decreases naturally ("decays") to safe levels. This interim stage spans years or decades, depending on the type of fuel. Most U.S. waste is currently stored in temporary storage sites requiring oversight, while suitable permanent disposal methods are discussed. As of 2003, the United States had accumulated about 49,000 metric tons of spent nuclear fuel from nuclear reactors. Underground storage at Yucca Mountain in U.S. has been proposed as permanent storage. After 10,000 years of radioactive decay, according to United States Environmental Protection Agency standards, the spent nuclear fuel will no longer pose a threat to public health and safety.

The amount of waste can be reduced in several ways, particularly reprocessing, as below. Even so, the remaining waste will be substantially radioactive for at least 300 years even if the actinides are removed, and for up to thousands of years if the actinides are left in. Even with separation of all actinides, and using fast breeder reactors to destroy by transmutation some of the longer-lived non-actinides as well, the waste must be segregated from the environment for one to a few hundred years, and therefore this is properly categorized as a long-term problem. Subcritical reactors or fusion reactors could also reduce the time the waste has to be stored. It has been argued that the best solution for the nuclear waste is above ground temporary storage since technology is rapidly changing. The current waste may well become a valuable resource in the future.

In countries with nuclear power, radioactive wastes comprise less than 1% of total industrial toxic wastes, which remain hazardous indefinitely unless they decompose or are treated so that they are less toxic or, ideally, completely non-toxic.

The nuclear industry also produces a volume of low-level radioactive waste in the form of contaminated items like clothing, hand tools, water purifier resins, and (upon decommissioning) the materials of which the reactor itself is built. In the United States, the Nuclear Regulatory Commission has repeatedly attempted to allow low-level materials to be handled as normal waste: landfilled, recycled into consumer items, etc. Most low-level waste releases very low levels of radioactivity and is only considered radioactive waste because of its history. For example, according to the standards of the NRC, the radiation released by coffee is enough to treat it as low level waste. Overall, nuclear power produces far less waste material than fossil-fuel based power plants. Coal-burning plants are particularly noted for producing large amounts of toxic and mildly radioactive ash due to concentrating naturally occurring metals and radioactive material from the coal.

Nuclear Reprocessing

Reprocessing can recover up to 95% of the remaining uranium and plutonium in spent nuclear fuel, putting it into new mixed oxide fuel. This also produces a reduction in long term radioactivity within the remaining waste, since this is largely short-lived fission products, and reduces its volume by over 90%. Reprocessing of civilian fuel from power reactors is currently done on large scale in Britain, France and (formerly) Russia, will be in China and perhaps India, and is being done on an expanding scale in Japan. Iran has announced its intention to complete the nuclear fuel cycle via reprocessing, a move which has led to criticism from the United States and the International Atomic Energy Agency. Unlike other countries, U.S. policy at one stage forbade recycling of used fuel; although this policy was reversed, spent fuel is all currently treated as waste.

Economy

Opponents of nuclear power argue that any of the environmental benefits are outweighed by safety compromises and by the costs related to construction and operation of nuclear power plants, including costs for spent-fuel disposal and plant retirement. Proponents of nuclear power respond that nuclear energy is the only power source which explicitly factors the estimated costs for waste containment and plant decommissioning into its overall cost, and that the quoted cost of fossil fuel plants is deceptively low for this reason. The cost of some renewables would be increased too if they included necessary back-up due to their intermittent nature.

A UK Royal Academy of Engineering report in 2004 looked at electricity generation costs from new plants in the UK. In particular it aimed to develop "a robust approach to compare directly the costs of intermittent generation with more dependable sources of generation". This meant adding the cost of standby capacity for wind, as well as carbon values up to £30 per tonne CO_2 for coal and gas. Wind power was calculated to be more than twice as expensive as nuclear power. Without a carbon tax, the cost of production through coal, nuclear and gas ranged £0.022-0.026/kWh and coal gasification was £0.032/kWh. When carbon tax was added (up to £0.025) coal came close to onshore wind (including back-up power) at £0.054/kWh-offshore wind is £0.072/kWh.

Nuclear power remained at £0.023/kWh either way, as it produces negligible amounts of CO_2. Nuclear figures included decommissioning costs.

In one study, certain gas cogeneration plants were calculated to be three to four times more cost-effective than nuclear power, if all the heat produced was used onsite or in a local heating system. However, the study estimated only 25 year plant lifetimes (60 is now common), 68% capacity factors were assumed (above 90% is now common), other conservatisms were applied, and nuclear power also produces heat which could be used in similar ways (although most nuclear power plants are located in remote areas). The study then found similar costs for nuclear power and most other forms of generation if not including external costs (such as back-up power).

Capital Costs

Generally, a nuclear power plant is significantly more expensive to build than an equivalent coal-fuelled or gas-fuelled plant. Coal is significantly more expensive than nuclear fuel, and natural gas significantly more expensive than coal-thus, capital costs aside, natural gas-generated power is the most expensive. However, servicing the capital costs for a nuclear power dominate the costs of nuclear-generated electricity, contributing about 70% of costs (assuming a 10% discount rate).

The recent liberalisation of the electricity market in many countries has made the economics of nuclear power generation less attractive. Previously a monopolistic provider could guarantee output requirements decades into the future. Private generating companies have to accept shorter output contracts and the risks of future competition, so desire a shorter return on investment period which favours generation plants with lower capital costs.

In many countries, licensing, inspection and certification of nuclear power plants has added delays and construction costs to their construction. In the U.S. many new regulations were put in place after the Three Mile Island partial meltdown. Gas-fired and coal-fired plants do not face such regulations. Because a power plant does not yield profits during construction, longer construction times translate directly into higher interest charges on borrowed construction funds. However, the regulatory processes for siting, licensing, and constructing have been standardized since their introduction, to make construction of newer and safer designs more attractive to companies.

In Japan and France, construction costs and delays are significantly diminished because of streamlined government licensing and certification procedures. In France, one model of reactor was type-certified, using a safety engineering process similar to the process used to certify aircraft models for safety. That is, rather than licensing individual reactors, the regulatory agency certified a particular design and its construction process to produce safe reactors. U.S. law permits type-licensing of reactors, a process which is about to be used.

To encourage development of nuclear power, under the Nuclear Power 2010 Program the U.S. Department of Energy (DOE) has offered interested parties the opportunity to introduce France's model for licensing and to subsidize 25% to 50% of the construction cost overruns due to delays for the first six new plants. Several applications were made, two sites have been chosen to receive new plants, and other projects are pending.

Operating Costs

In general, coal and nuclear plants have the same types of operating costs (operations and maintenance plus fuel costs). However, nuclear and coal differ in the relative size of those costs. Nuclear has lower fuel costs but higher operating and maintenance costs. In recent times in the United States savings due to lower fuel cost have not been low enough for nuclear to repay its higher investment cost. Thus no new nuclear reactors have been ordered in the United States since the 1970s. Coal's operating cost advantages have only rarely been sufficient to encourage the construction of new coal based power generation. Around 90 to 95 percent of new power plant construction in the United States has been natural gas-fired.

To be competitive in the current market, both the nuclear and coal industries must reduce new plant investment costs and construction time. The burden is clearly greater for nuclear producers than for coal producers, because investment costs are higher for nuclear plants. Operation and maintenance costs are particularly important because they represent a large portion of costs for nuclear power.

One of the primary reasons for the uncompetitiveness of the U.S. nuclear industry has been the lack of any measure that provides an economic incentive to reduce carbon emissions (carbon tax). Many economists and environmentalists have called for a mechanism to take into account the negative externalities of coal and gas consumption. In such an environment, it is argued that nuclear will become cost-competitive in the United States.

Subsidies

Critics of nuclear power claim that it is the beneficiary of inappropriately large economic subsidies-mainly taking the forms of taxpayer-funded research and development and limitations on disaster liability-and that these subsidies, being subtle and indirect, are often overlooked when comparing the economics of nuclear against other forms of power generation. However, competing energy sources also receive subsidies. Fossil fuels receive large direct and indirect subsidies, such as tax benefits and not having to pay for the Greenhouse gases they emit. Renewables receive large direct production subsidies and tax breaks in many nations.

Energy research and development (R&D) for nuclear power has and continues to receive much larger state subsidies than R&D for renewable energy or fossil fuels. However, today most of this takes places in Japan and France: in most other nations renewable R&D get more money. In the U.S., public research money for nuclear fission declined from 2,179 to 35 million dollars between 1980 and 2000. However, in order to restart the industry, the next six U.S. reactors will receive subsidies equal to those of renewables and, in the event of cost overruns due to delays, at least partial compensation for the overruns.

According to the DOE, insurance for nuclear or radiological incidents in the U.S., is subsidized by the Price-Anderson Nuclear Industries Indemnity Act. In July 2005, Congress extended this Act to newer facilities. In the UK, the Nuclear Installations Act of 1965 governs liability for nuclear damage for which a UK nuclear licensee is responsible. The Vienna Convention on Civil Liability for Nuclear Damage puts in place an international framework for nuclear liability.

Other Economic Issues

Nuclear Power plants tend to be most competitive in areas where other fuel resources are not readily available-France, most notably, has almost no native supplies of fossil fuels. The province of Ontario, Canada is already using all of its best sites for hydroelectric power, and has minimal supplies of fossil fuels, so a number of nuclear plants have been built there. India is also building new nuclear plants to supplement its vast coal reserves and coal-generated electricity. Conversely, in the United Kingdom, according to the government's Department Of Trade And Industry, no further nuclear power stations are to be built, due to the high cost per unit of nuclear power compared to fossil fuels. However, the British government's chief scientific advisor David King reports that building one more generation of nuclear power plants may be necessary. China tops the list of planned new plants, due to its rapidly expanding economy and fervent construction in many types of energy projects.

Most new gas-fired plants are intended for peak supply. The larger nuclear and coal plants cannot quickly adjust their instantaneous power production, and are generally intended for baseline supply. The market price for baseline power has not increased as rapidly as that for peak demand. Some new experimental reactors, notably pebble bed modular reactors, are specifically designed for peaking power.

Any effort to construct a new nuclear facility around the world, whether an older design or a newer experimental design, must deal with NIMBY objections. Given the high profile of both the Three Mile Island and Chernobyl accidents, few municipalities welcome a new nuclear reactor, processing plant, transportation route, or experimental nuclear burial ground within their borders, and many have issued local ordinances prohibiting the development of nuclear power. However, a few U.S. areas with nuclear units are campaigning for more.

Current nuclear reactors return around 40-60 times the invested energy when using life cycle analysis. This is better than coal, natural gas, and current renewables except hydropower.

The Rocky Mountain Institute gives other reasons why nuclear power plants may not be economical. In the U.S. this includes long lead times on risky investments, and the more cost-effective approach of investing in efficiency instead of new power plants.

Nuclear power, coal, and wind power are currently the only realistic large scale energy sources that would be able to replace oil and natural gas after a peak in global oil and gas production has been reached. However, The Rocky Mountain Institute claims that in the U.S. increases in transportation efficiency and stronger, lighter cars would replace most oil usage with what it calls negawatts. Biofuels can then substitute for a significant portion of the remaining oil use. Efficiency, insulation, solar thermal, and solar photovoltaic technologies can substitute for most natural gas usage after a peak in production.

Nuclear proponents often assert that renewable sources of power have not solved problems like intermittent output, high costs, and diffuse output which requires the use of large surface areas and much construction material and which increases distribution losses. For example, studies in Britain have shown that increasing wind power production contribution to 20% of all energy production, without costly pumped hydro or electrolysis/fuel cell storage, would only reduce coal or nuclear power plant capacity by 6.7% (from 59 to 55 GWe) since they must remain as backup in the absence of power storage. Nuclear proponents often claim that increasing the contribution of intermittent energy sources above that is not possible with current technology. Some renewable energy sources, such as solar, overlap well with peak electrical production and reduce the need of spare generating capacity. Future applications that use electricity when it is available (e.g. for pressurizing water systems, desalination, or hydrogen generation) would help to reduce the spare generation capacity required by both nuclear and renewable energy sources.

CONCERNS ABOUT NUCLEAR POWER

The Ignalina nuclear power plant in Lithuania contains two RBMK reactors. Because of the safety flaws of the design, the closure of the plant was a condition of Lithuania's entry into the EU. The first of the two reactors was closed down in 2004 and the second is scheduled for shutdown by 2009. (Photograph courtesy of the Nordic Council).

Opponents argue that a major disadvantage of the use of nuclear reactors is the threat of a nuclear accident or terrorist attack and the possible resulting exposure to radiation.

Proponents argue that the potential for a meltdown in non-Russian-designed reactors is very small due to the care taken in designing adequate safety systems, and that the nuclear industry has much better statistics regarding humans deaths from occupational accidents than coal or hydropower. While the Chernobyl accident caused great negative health, economic, environmental and psychological effects in a widespread area, the accident at Chernobyl was caused by a combination of the faulty RBMK reactor design, the lack of a properly designed containment building, poorly trained operators, and a non-existent safety culture. The RBMK design, unlike nearly all designs used in the Western world, featured a positive void coefficient, meaning that a malfunction could result in ever-increasing generation of heat and radiation until the reactor was breached. Even at Three Mile Island, the most severe civilian nuclear accident in the non-Soviet world, the reactor vessel and containment building were never breached, even though it had suffered a core meltdown, so that very little radiation (well below natural background radiation levels) was released into the environment.

Design changes are being pursued in the hope of lessening some of the risks of fission reactors; in particular, automated and passively safe designs are being pursued. Fusion reactors which may come to exist in the future theoretically have little risk since the fuel contained in the reaction chamber is only enough to sustain the reaction for about a minute, whereas a fission reactor contains about a year's supply of fuel. Subcritical reactors never have a self sustained nuclear chain reaction.

Opponents of nuclear power express concerns that nuclear waste is not well protected, and that it can be released in the event of terrorist attack, quoting a 1999 Russian incident where workers were caught trying to sell 5 grams of radioactive material on the open market, or the incident in 1993 where

Russian workers were caught selling 4.5 kilograms of enriched uranium. The UN has since called upon world leaders to improve security in order to prevent radioactive material falling into the hands of terrorists, sometimes leading to the guarding of nuclear shipments by thousands of police. Other energy sources, such as hydropower plants and LNG carriers, are more vulnerable to accidents and attacks. Proponents of nuclear power contend, however, that nuclear waste is already well protected, and state their argument that there has been no accident involving any form of nuclear waste from a civilian program worldwide. In addition, they point to large studies carried out by NRC and other agencies that tested the robustness of both reactor and waste fuel storage, and found that they should be able to sustain a terrorist attack comparable to the September 11 terrorist attacks. Spent fuel is usually housed inside the plant's "protected zone" or a spent nuclear fuel shipping cask; stealing it for use in a "dirty bomb" is extremely difficult. Exposure to the intense radiation would almost certainly quickly incapacitate or kill any terrorists who attempt to do so.

According to the Nuclear Regulatory Commission, 20 American states have requested stocks of potassium iodide which the NRC suggests should be available for those living within 10 miles of a nuclear power plant in the unlikely event of a severe accident.

HEALTH EFFECTS ON POPULATIONS

Most of the human exposure to radiation comes from natural background radiation. Most of the remaining exposure comes from medical procedures. Several large studies in the U.S., Canada, and Europe have found no evidence of any increase in cancer mortality among people living near nuclear facilities. For example, in 1990, the National Cancer Institute (NCI) of the National Institutes of Health announced that, after doing a large-scale study which evaluated the mortality rates from 16 types of cancer, no increased incidence of cancer mortality was found for people living near 62 nuclear installations in the United States. The study also showed no increase in the incidence of childhood leukemia mortality in the study of surrounding counties after the start-up of the nuclear facilities.

The NCI study, the broadest of its kind ever conducted, surveyed 900,000 cancer deaths in counties near nuclear facilities.

Aside from the immediate effects of the Chernobyl accident, there is continuing impact from soils containing radioactivity in Ukraine and Belarus. For this reason a Zone of alienation was established around the Chernobyl plant.

In March, 2006, safety reviews found that several nuclear plants in the United States have been leaking water contaminated with tritium into the ground. (The discharges were intended to go through discharge pipes into rivers, at levels which would be below regulatory limits. However, by leaking into the ground, very low levels of tritium reached drinking water supplies.) The attorney general of Illinois announced that she was filing a lawsuit against Exelon because of six such leaks, demanding that the utility provide substitute water supplies to residents although no well outside company property shows levels that exceed drinking water standards. According to the NRC, "The inspection determined that public health and safety has not been adversely affected and the dose consequence to the public that can be attributed to current onsite conditions is negligible with respect to NRC regulatory limits." However, the chairman of the Nuclear Regulatory Commission, said, "They're going to have to fix it."

NUCLEAR PROLIFERATION

Opponents of nuclear power point out that nuclear technology is often dual-use, and much of the same materials and knowledge used in a civilian nuclear program can be used to develop nuclear weapons. This concern is known as nuclear proliferation and is a major reactor design criterion.

The military and civil purposes for nuclear energy are intertwined in most countries with nuclear capabilities. In the U.S., for example, the first goal of the Department of Energy is "to advance the national, economic, and energy security of the United States; to promote scientific and technological innovation in support of that mission; and to ensure the environmental cleanup of the national nuclear weapons complex."

The enriched uranium used in most nuclear reactors is not concentrated enough to build a bomb. Most nuclear reactors run on 4% enriched uranium; Little Boy used 80% enriched uranium; while lower enrichment levels could be used, the minimum bomb size would rapidly become infeasibly large as the level was decreased. However, the same plants and technology used to enrich uranium for power generation can be used to make the highly enriched uranium needed to build a bomb.

In addition, the plutonium produced in power reactors, if concentrated through reprocessing, can be used for a bomb. While the plutonium resulting from normal reactor fuelling cycles is less than ideal for weapons use because of the concentration of Pu-240, a usable weapon can be produced from it. If the reactor is operated on very short fuelling cycles, bomb-grade plutonium can be produced. However, such operation would be virtually impossible to camouflage in many reactor designs, as the frequent shutdowns for refuelling would be obvious, for instance in satellite photographs.

It is widely believed that the nuclear programs of India and Pakistan used CANDU reactors to produce fissionable materials for their weapons; however, this is not entirely accurate. Both Canada (by supplying the 40 MW research reactor) and the United States (by supplying 21 tons of heavy water) supplied India with the technology necessary to create a nuclear weapons programme, dubbed CIRUS (Canada-India Reactor, United States). Because international policies did not dictate usage of nuclear technology transfers, India was able to use the technology to create a nuclear weapon. Pakistan is believed to have produced the material for its weapons from an indigenous enrichment program.

To prevent weapons proliferation, safeguards on nuclear technology were published in the Nuclear Non-Proliferation Treaty (NPT) and monitored since 1968 by the International Atomic Energy Agency (IAEA). Nations signing the treaty are required to report to the IAEA what nuclear materials they hold and their location. They agree to accept visits by IAEA auditors and inspectors to verify independently their material reports and physically inspect the nuclear materials concerned to confirm physical inventories of them in exchange for access to nuclear materials and equipment on the global market.

Several states did not sign the treaty and were able to use international nuclear technology (often procured for civilian purposes) to develop nuclear weapons (India, Pakistan, Israel, and South Africa). South Africa has since signed the NPT, and now holds the distinction of being the only known state to have indigenously produced nuclear weapons, and then verifiably dismantled them. Of those who have signed the treaty and received shipments of nuclear paraphernalia, many states have either claimed to, or been accused of, attempting to use supposedly civilian nuclear power plants for developing weapons, including Iran and North Korea.

Certain types of reactors are more conducive to producing nuclear weapons materials than others, and a number of international disputes over proliferation have centered on the specific model of reactor being contracted for in a country suspected of nuclear weapon ambitions.

New technology, like SSTAR, may lessen the risk of nuclear proliferation by providing sealed reactors with a limited self-contained fuel supply and with restrictions against tampering.

One possible obstacle for expanding the use of nuclear power might be a limited supply of uranium ore, without which it would become necessary to build and operate breeder reactors. However, at current usage there is sufficient uranium for an extended period-"In summary, the actual recoverable uranium supply is likely to be enough to last several hundred (up to 1000) years, even using standard reactors." Breeder reactors have been banned in the U.S. since President Jimmy Carter's administration prohibited reprocessing because of what it regarded as the unacceptable risk of proliferation of weapons-grade materials.

Some proponents of nuclear power agree that the risk of nuclear proliferation may be a reason to prevent nondemocratic developing nations from gaining any nuclear technology but argue that this is no reason for democratic developed nations to abandon their nuclear power plants, especially since it seems that democracies refrain from war against each other.

Proponents also note that nuclear power, like some other power sources, provides steady energy at a consistent price without competing for energy resources from other countries, something that may contribute to wars.

In February, 2006, a new U.S. initiative, the Global Nuclear Energy Partnership was announced. It would be an international effort to reprocess fuel in a manner making proliferation infeasible, while making nuclear power available to developing countries.

ENVIRONMENTAL EFFECTS

Non-radioactive water vapour is the significant operating emission from nuclear power plants. Fission produces gases such as iodine-131 or Xenon-133. These primarily remain within the fuel rods, but with some postulated fuel failure, small amounts of the gases can be released in to the reactor coolant. The chemical control systems isolate the radioactive gases which have to be stored on-site for several half-lives until they have decayed to safe levels. Iodine-131 and Xenon-133 have halflives of 8.0 and 5.2 days respectively, and thus have to be stored for a few months to decay to safe levels.

Nuclear generation does not directly produce sulfur dioxide, nitrogen oxides, mercury or other pollutants associated with the combustion of fossil fuels (pollution from fossil fuels is blamed for many deaths each year in the U.S. alone). It also does not directly produce carbon dioxide, which has led some environmentalists to advocate increased reliance on nuclear energy as a means to reduce greenhouse gas emissions (which contribute to global warming).

Like any power source (including renewables like wind and solar energy), the facilities to produce and distribute the electricity require energy to build and subsequently decommission. Mineral ores must be collected and processed to produce nuclear fuel. These processes are either directly powered by diesel and gasoline engines, or draw electricity from the power grid, which may be generated from fossil fuels. Life cycle analyses assess the amount of energy consumed by these processes (given today's mix of energy resources) and calculate, over the lifetime of a nuclear power plant, the amount of carbon dioxide saved (related to the amount of electricity produced by the plant) vs. the amount of carbon dioxide used (related to construction and fuel acquisition).

Several life cycle analyses show similar emissions per kilowatt-hour from nuclear power and from renewables such as wind power. According to one life cycle study by van Leeuwen and Smith from 2001-2005, carbon dioxide emissions from nuclear power per kilowatt hour could range from 20% to 120% of those for natural gas-fired power stations depending on the availability of high grade ores. The study was critiqued by World Nuclear Association (WNA), rebutted in 2003, then dismissed by the WNA in 2006 based on its own life-cycle-energy calculation (with comparisons).

In 2006, a UK government advisory panel, The Sustainable Development Commission, concluded that if the UK's existing nuclear capacity were doubled, it would provide an 8% decrease in total UK CO2 emissions by 2035. This can be compared to the country's goal to reduce greenhouse gas emissions by 60% by 2050. As of 2006, the UK government was to publish its official findings later in the year.

Waste Heat in Water Systems

Nuclear reactors require cooling, typically with water (sometimes indirectly). The process of using water to extract energy from a heat source requires a cooling source, this process is described by the Rankine cycle. There is a limitation on the amount of heat that can be converted into energy through the Rankine Cycle. The excess heat must be rejected as waste heat, this is where the cooling water is required. Rivers are the most common source of cooling water, as well as the destination for waste heat. The temperature of exhaust water must be regulated to avoid killing fish; long-term impact of hotter-than-natural water on ecosystems is an environmental concern. In most new facilities, this problem is solved by using cooling towers. This is true of all traditional power plant designs, including coal, oil, and natural gas plants, which also rely on the Rankine cycle to produce their energy. All four types of plants differ in their heat source, be it nuclear fission or burning fossil fuels.

The need to regulate exhaust temperature can limit generation capacity. On extremely hot days, which is when demand can be at its highest, the capacity of a nuclear plant may go down because the incoming water is warmer to begin with and is thus less effective as a coolant, per unit volume. This was a significant factor during the European heat wave of 2003. Engineers consider this in making improved power plant designs because increased cooling capacity will increase capital costs. The global increase in average temperature has required some plants in the southeast United States to revise their technical specifications to allow operation with their cooling water sources at higher temperatures.

PETROLEUM

Petroleum (from Latin petra-rock and Greek oleum-oil) or crude oil is a naturally occurring liquid found in formations in the Earth consisting of a complex mixture of hydrocarbons (mostly alkanes) of various lengths. In its naturally occurring form, it may contain other nonmetallic elements such as sulfur, oxygen, and nitrogen. It is usually black or dark brown (although it may be yellowish or even greenish) but varies greatly in appearance, depending on its composition.

Petroleum is used mostly, by volume, for producing fuel oil and petrol (gasoline), both important "primary energy" sources. 84% (37 of 42 gallons in a typical barrel) of the hydrocarbons present in petroleum is converted into energy-rich fuels, including gasoline, diesel, jet, heating, and other fuel oils, and liquefied petroleum gas; Due to its incredible energy density and relative abundance, it has become the world's most important source of energy since the mid-1950s. Petroleum is also the raw material for many chemical products, including solvents, fertilizers, pesticides, and plastics; the 16% not used for energy production is converted into these other materials.

Petroleum is found in porous rock formations in the upper strata of some areas of the Earth's crust. Known reserves of petroleum are typically estimated at around 1.2 trillion barrels with at least one estimate as high as 3.74 trillion barrels. Consumption is currently around 84 million barrels per day, or 31 billion barrels per year. Because of pumping difficulties, usable oil reserves are only about 1/3 of total reserves. At current consumption levels, world oil supply would be gone in about 32 years, around 2039. However, this ignores any additions to known reserves, changes in demand, etc. As the supply of petroleum becomes more scarce, consumers may look to energy development such as solar power, wind power, ethanol, photovoltaic, nuclear power, hydrogen (which combusts according to $2H_2(g) + O_2(g) \rightarrow 2H_2O(l) + 572$ kJ), or oil from oil shale, and tar sands. Petroleum forms naturally within the earth too slowly to be sustainable for human use.

BIOGENIC THEORY

Most geologists view crude oil and natural gas as the product of compression and heating of ancient organic materials over geological time. According to this theory, oil is formed from the preserved remains of prehistoric zooplankton and algae which have been settled to the sea bottom in large quantities under anoxic conditions. Terrestrial plants, on the other hand, tend to form coal. Over geological time this organic matter, mixed with mud, is buried under heavy layers of sediment. The resulting high levels of heat and pressure cause the remains to metamorphose, first into a waxy material known as kerogen which is found in various oil shales around the world, and then with more heat into liquid and gaseous hydrocarbons in a process known as catagenesis. Because most hydrocarbons are lighter than rock or water, these sometimes migrate upward through adjacent rock layers until they become trapped beneath impermeable rocks, within porous rocks called reservoirs.

Concentration of hydrocarbons in a trap forms an oil field, from which the liquid can be extracted by drilling and pumping. Geologists often refer to an "oil window" which is the temperature range that oil forms in-below the minimum temperature oil remains trapped in the form of kerogen, and above the maximum temperature the oil is converted to natural gas through the process of thermal cracking. Though this happens at different depths in different locations around the world, a 'typical' depth for the oil window might be 4-6 km. Note that even if oil is formed at extreme depths, it may be trapped at much shallower depths, even if it is not formed there (the

Athabasca Oil Sands is one example). Three conditions must be present for oil reservoirs to form: first, a source rock rich in organic material buried deep enough for subterranean heat to cook it into oil; second, a porous and permeable reservoir rock for it to accumulate in; and last a cap rock (seal) that prevents it from escaping to the surface.

If an oil well were to run dry and be capped, it would likely fill back to its original supply eventually. There is considerable question about how long this would take. Some formations appear to have a regeneration time of decades. Majority opinion is that oil is being formed at less than 1% of the current consumption rate.

The vast majority of oil that has been produced by the earth has long ago escaped to the surface and been biodegraded by oil-eating bacteria. Oil companies are looking for the small fraction that has been trapped by this rare combination of circumstances. Oil sands are reservoirs of partially biodegraded oil still in the process of escaping, but contain so much migrating oil that, although most of it has escaped, vast amounts are still present-more than can be found in conventional oil reservoirs. On the other hand, oil shales are source rocks that have never been buried deep enough to convert their trapped kerogen into oil. "Oil is a reneweable resource." Quoted from the scientist, Nandu Krishnan.

The reactions that produce oil and natural gas are often modeled as first order breakdown reactions, where kerogen is broken down to oil and natural gas by a set of parallel reactions, and oil eventually breaks down to natural gas by another set of reactions. The first set was originally patented in 1694 under British Crown Patent No. 330 covering, "a way to extract and make great quantities of pitch, tarr, and oyle out of a sort of stone."

The latter set is regularly used in petrochemical plants and oil refineries.

Abiogenic Theory

The idea of abiogenic petroleum origin was championed in the Western world by astronomer Thomas Gold based on thoughts from Russia, mainly on studies of Nikolai Kudryavtsev. The idea proposes that hydrocarbons of purely geological origin exist in the planet. Hydrocarbons are less dense than aqueous pore fluids, and are proposed to migrate upward through deep fracture networks. Thermophilic, rock-dwelling microbial life-forms are proposed to be in part responsible for the biomarkers found in petroleum.

This theory is a minority opinion, especially amongst geologists; no oil companies are currently known to explore for oil based on this theory.

Classification

The oil industry classifies "crude" by the location of its origin (e.g., "West Texas Intermediate, WTI" or "Brent") and often by its relative weight or viscosity ("light", "intermediate" or "heavy"); refiners may also refer to it as "sweet," which means it contains relatively little sulfur, or as "sour," which means it contains substantial amounts of sulfur and requires more refining in order to meet current product specifications. Each crude oil has unique molecular characteristics which are understood by the use of crude oil assay analysis in petroleum laboratories.

Barrels from an area in which the crude oil's molecular characteristics have been determined and the oil has been classified are used as pricing references throughout the world. These references are known as Crude oil benchmarks and are listed;

- Brent Crude, comprising 15 oils from fields in the Brent and Ninian systems in the East Shetland Basin of the North Sea. The oil is landed at Sullom Voe terminal in the Shetlands. Oil production from Europe, Africa and Middle Eastern oil flowing West tends to be priced off the price of this oil, which forms a benchmark.
- West Texas Intermediate (WTI) for North American oil.
- Dubai, used as benchmark for Middle East oil flowing to the Asia-Pacific region.
- Tapis (from Malaysia, used as a reference for light Far East oil)

- Minas (from Indonesia, used as a reference for heavy Far East oil)
- The OPEC Reference Basket, a weighted average of oil blends from various OPEC countries.

MEANS OF PRODUCTION

The most common method of obtaining petroleum is extracting it from oil wells found in oil fields. After the well has been located, various methods are used to recover the petroleum. Primary recovery methods are used to extract oil that is brought to the surface by underground pressure, and can generally recover about 20% of the oil present. After the oil pressure has depleted to the point that the oil is no longer brought to the surface, secondary recovery methods draw another 5 to 10% of the oil in the well to the surface. Finally, when secondary oil recovery methods are no longer viable, tertiary recovery methods reduce the viscosity of the oil in order to bring more to the surface.

Alternative Methods

As oil prices continue to escalate, other alternatives to producing oil have been gaining importance. The best known such methods involve extracting oil from sources such as oil shale or tar sands. These resources are known to exist in large quantities; however, extracting the oil at low cost without negatively impacting the environment remains a challenge.

It is also possible to transform natural gas or coal into oil (or, more precisely, the various hydrocarbons found in oil). The best-known such method is the Fischer-Tropsch process. It was a concept pioneered in Nazi Germany when imports of petroleum were restricted due to war and Germany found a method to extract oil from coal. It was known as Ersatz ("substitute" in German), and accounted for nearly half the total oil used in WWII by Germany. However, the process was used only as a last resort as naturally occurring oil was much cheaper. As crude oil prices increase, the cost of coal to oil conversion becomes comparatively cheaper. The method involves converting high ash coal into synthetic oil in a multi-stage process. Ideally, a ton of coal produces nearly 200 liters (1.25 bbl, 52 US gallons) of crude, with by-products ranging from tar to rare chemicals. Currently, two companies have commercialised their Fischer-Tropsch technology. Shell in Bintulu, Malaysia, uses natural gas as a feedstock, and produces primarily low-sulfur diesel fuels. Sasol in South Africa uses coal as a feedstock, and produces a variety of synthetic petroleum products.

The process is today used in South Africa to produce most of the country's diesel fuel from coal by the company Sasol. The process was used in South Africa to meet its energy needs during its isolation under Apartheid. This process has received renewed attention in the quest to produce low sulfur diesel fuel in order to minimize the environmental impact from the use of diesel engines.

An alternative method of converting coal into petroleum is the Karrick process, which was pioneered in the 1930s in the United States. It uses high temperatures in the absence of ambient air, to distill the short-chain hydrocarbons of petroleum out of coal.

More recently explored is thermal depolymerization (TDP), a process for the reduction of complex organic materials into light crude oil. Using pressure and heat, long chain polymers of hydrogen, oxygen, and carbon decompose into short-chain petroleum hydrocarbons. This mimics the natural geological processes thought to be involved in the production of fossil fuels. In theory, TDP can convert any organic waste into petroleum.

History

Petroleum, in some form or other, is not a substance new in the world's history. More than four thousand years ago, according to Herodotus and confirmed by Diodorus Siculus, asphalt was employed in the construction of the walls and towers of Babylon; there were oil pits near Ardericca (near Babylon), and a pitch spring on Zacynthus. Great quantities of it were found on the banks of the river Issues, one of the tributaries of the Euphrates. Ancient Persian tablets indicate the medicinal and lighting uses of petroleum in the upper levels of their society.

The first oil wells were drilled in China in the 4th century or earlier. They had depths of up to 243 meters and were drilled using bits attached to bamboo poles. The oil was burned to evaporate brine and produce salt. By the 10th century, extensive bamboo pipelines connected oil wells with salt springs. The ancient records of China and Japan are said to contain many allusions to the use of natural gas for lighting and heating. Petroleum was known as burning water in Japan in the 7th century.

In the 8th century, the streets of the newly constructed Baghdad were paved with tar, derived from easily accessible petroleum from natural fields in the region. In the 9th century, oil fields were exploited in the area around modern Baku, Azerbaijan, to produce naphtha. These fields were described by the geographer Masudi in the 10th century, and by Marco Polo in the 13th century, who described the output of those wells as hundreds of shiploads.

The earliest mention of American petroleum occurs in Sir Walter Raleigh's account of the Trinidad Pitch Lake in 1595; whilst thirty-seven years later, the account of a visit of a Franciscan, Joseph de la Roche d'Allion, to the oil springs of New York was published in Sagard's Histoire du Canada. A Russian traveller, Peter Kalm, in his work on America, published in 1748, showed on a map the oil springs of Pennsylvania.

The modern history of petroleum began in 1846, with the discovery of the process of refining kerosene from coal by Atlantic Canada's Abraham Pineo Gesner. The first modern oil well was drilled in 1848 by Russian engineer F.N. Semyonov, on the Apsheron Peninsula north-east of Baku. Poland's Ignacy Lukasiewicz discovered a means of refining kerosene from the more readily available "rock oil" ("petr-oleum") in 1852 and the first rock oil mine was built in Bóbrka, near Krosno in southern Poland in the following year. These discoveries rapidly spread around the world, and Meerzoeff built the first Russian refinery in the mature oil fields at Baku in 1861. At that time Baku produced about 90% of the world's oil. The battle of Stalingrad was fought over Baku (now the capital of the Azerbaijan Republic).

The first commercial oil well drilled in North America was in Oil Springs, Ontario, Canada in 1858, dug by James Miller Williams. The American petroleum industry began with Edwin Drake's drilling of a 69-foot-deep oil well in 1859, on Oil Creek near Titusville, Pennsylvania, for the Seneca Oil Company (originally yielding 25 barrels a day, by the end of the year output was at the rate of 15 barrels). The industry grew slowly in the 1800s, driven by the demand for kerosene and oil lamps. It became a major national concern in the early part of the 20th century; the introduction of the internal combustion engine provided a demand that has largely sustained the industry to this day. Early "local" finds like those in Pennsylvania and Ontario were quickly exhausted, leading to "oil booms" in Texas, Oklahoma, and California.

Early production of crude petroleum in the United States:

- 1859: 2,000 barrels
- 1869: 4,215,000 barrels
- 1879: 19,914,146 barrels
- 1889: 35,163,513 barrels
- 1899: 57,084,428 barrels
- 1906: 126,493,936 barrels

By 1910, significant oil fields had been discovered in Canada (specifically, in the province of Alberta), the Dutch East Indies (1885, in Sumatra), Iran (1908, in Masjed Soleiman), Peru, Venezuela, and Mexico, and were being developed at an industrial level.

Even until the mid-1950s, coal was still the world's foremost fuel, but oil quickly took over. Following the 1973 energy crisis and the 1979 energy crisis, there was significant media coverage of oil supply levels. This brought to light the concern that oil is a limited resource that will eventually run out, at least as an economically viable energy source. At the time, the most common and popular predictions were always quite dire, and when they did not come true, many dismissed all such discussion. The future of petroleum as a fuel remains somewhat controversial. USA Today news (2004) reports that there are 40 years of petroleum left in the ground. Some would argue that because the total amount of petroleum is finite, the dire predictions of the 1970s have merely been postponed. Others argue that technology will continue to allow for the production of cheap hydrocarbons and that the earth has vast sources of unconventional petroleum reserves in the form of tar sands, bitumen fields and oil shale that will allow for petroleum use to continue in the future, with both the Canadian tar sands and United States shale oil deposits representing potential reserves matching existing liquid petroleum deposits worldwide.

Today, about 90% of vehicular fuel needs are met by oil. Petroleum also makes up 40% of total energy consumption in the United States, but is responsible for only 2% of electricity generation. Petroleum's worth as a portable, dense energy source powering the vast majority of vehicles and as the base of many industrial chemicals makes it one of the world's most important commodities. Access to it was a major factor in several military conflicts, including World War I, World War II and the Persian Gulf War.

The top three oil producing countries are Saudi Arabia, Russia, and the United States. About 80% of the world's readily accessible reserves are located in the Middle East, with 62.5% coming from the Arab 5: Saudi Arabia (12.5%), UAE, Iraq, Qatar and Kuwait. However, with today's oil prices, Venezuela has larger reserves than Saudi Arabia due to nonconventional crude reserves derived from bitumen.

Uses

The chemical structure of petroleum is composed of hydrocarbon chains of different lengths. Because of this, petroleum may be taken to oil refineries and the hydrocarbon chemicals separated by distillation, to be used for a variety of purposes.

Fuels

- Ethane and other short-chain alkanes which are used as fuel
- Diesel fuel
- Fuel oils

- Gasoline
- Jet fuel
- Kerosene
- Liquid petroleum gas (LPG)

Other Derivatives

Certain types of resultant hydrocarbons may be mixed with other non-hydrocarbons, to create other end products:

- Alkenes which can be manufactured into plastics
- Lubricants (produces light machine oils, motor oils, and greases, adding viscosity stabilizers as required).
- Wax, used in the packaging of frozen foods, among others.
- Sulfuric acid finishing and shipping. This is a useful industrial material, usually prepared as the acid precursor oleum, a by-product of sulfur removal from fuels.
- Bulk tar.
- Asphalt
- Petroleum coke, used in speciality carbon products or as solid fuel.
- Paraffin wax

ALTERNATIVES TO PETROLEUM FUELS (GASOLINE AND DIESEL)

The term alternative propulsion or "alternative methods of propulsion" includes both:

- alternative fuels used in standard or modified internal combustion engines (i.e. combustion hydrogen).
- propulsion systems not based on internal combustion, such as those based on electricity (for example, electric or hybrid vehicles), compressed air, or fuel cells (i.e. hydrogen fuel cells).

Nowadays, cars can be classified between the next main groups:

- Parapetro cars, this is, only can use petroleum, biodiesel and biobutanol (instead of gasoline).
- Hybrid vehicle, that uses petroleum and other source, generally, electricity.
- Petrofree car, that can not use petroleum, like 100 % electric cars, hydrogen vehicles...

Environmental Effects

The presence of oil has significant social and environmental impacts, from accidents and routine activities such as seismic exploration, drilling, and generation of polluting wastes. Oil extraction is costly and sometimes environmentally damaging, although Dr. John Hunt of the Woods Hole Oceanographic Institution pointed out in a 1981 paper that over 70% of the reserves in the world are associated with visible macroseepages, and many oil fields are found due to natural leaks. Offshore exploration and extraction of oil disturbs the surrounding marine environment.

But at the same time, offshore oil platforms also form micro-habitats for marine creatures. Extraction may involve dredging, which stirs up the seabed, killing the sea plants that marine creatures need to survive. Crude oil and refined fuel spills from tanker ship accidents have damaged ecosystems in Alaska, the Galapagos Islands, Spain, and many other places.

Burning oil releases carbon dioxide into the atmosphere, which contributes to global warming. Per energy unit, oil produces less CO_2 than coal, but more than natural gas. However, oil's unique role as a transportation fuel makes reducing its CO_2 emissions a particularly thorny problem; amelioration strategies such as carbon sequestering are generally geared for large power plants, not individual vehicles.

Renewable energy alternatives do exist, but given current technology, alternatives are uneconomical. Solar, wind, geothermal, and other renewable electricity sources cannot directly replace high energy density liquid petroleum for transportation use; instead automobiles and other equipment must be altered to allow using electricity (in batteries) or hydrogen (via fuel cells or internal combustion) which can be produced from renewable sources. Other options include using biomass-origin liquid fuels (ethanol, biodiesel). Any combination of solutions to replace petroleum as a liquid transportation fuel will be a very large undertaking.

FUTURE OF OIL

The Hubbert peak theory (also known as peak oil) is a proposition which predicts that future world petroleum production must inevitably reach a peak and then decline as these reserves are exhausted. It also suggests a method to calculate mathematically the timing of this peak, based on past production rates, past discovery rates, and proven oil reserves.

Controversy surrounds the theory for numerous reasons. Past predictions regarding the timing of the global peak have failed, causing a number of observers to disregard the theory. Further, predictions regarding the timing of the peak are highly dependent on the past production and discovery data used in the calculation.

Proponents of peak oil theory also refer as an example of their theory, that when any given oil well produces oil in similar volumes to the amount of water used to obtain the oil, it tends to produce less oil afterwards, leading to the relatively quick exhaustion and/or commercial inviability of the well in question.

The issue can be considered from the point of view of individual regions or of the world as a whole. Hubbert's prediction for when US oil production would peak turned out to be correct, and after this occurred in 1971-causing the US to lose its excess production capacity-OPEC was finally able to manipulate oil prices, which led to the 1973 oil crisis. Since then, most other countries have also peaked: the United Kingdom's North Sea, for example in the late 1990s.

China has confirmed that two of its largest producing regions are in decline, and Mexico's national oil company, Pemex, has announced that Cantarell Field, one of the world's largest offshore fields, is expected to peak in 2006, and then decline 14% per annum.

It is difficult to predict the oil peak in any given region (due to the lack of transparency in accounting of global oil reserves). Based on available production data, proponents have previously (and incorrectly) predicted the peak for the world to be in years 1989, 1995, or 1995-2000.

However these predictions date from before the recession of the early 1980s, and the consequent reduction in global consumption, the effect of which was to delay the date of any peak by several years. A new prediction by Goldman Sachs picks 2007 for oil and some time later for natural gas. Just as the 1971 U.S. peak in oil production was only clearly recognized after the fact, a peak in world production will be difficult to discern until production clearly drops off.

Many proponents of the Hubbert peak theory expound the belief that the production peak is imminent, for various reasons. 2005 saw a dramatic fall in announced new oil projects coming to production from 2008 onwards-in order to avoid the peak, these new projects would have to not only make up for the depletion of current fields, but increase total production annually to meet increasing demand.

2005 also saw substantial increases in oil prices due to a number of circumstances, including war and political instability. Oil prices rose to new highs. Analysts such as Kenneth Deffeyes argue that these price increases indicate a general lack of spare capacity, and the price fluctuations can be interpreted as a sign that peak oil is imminent.

Costing

References to the oil prices are usually either references to the spot price of either WTI/Light Crude as traded on New York Mercantile Exchange (NYMEX) for delivery in Cushing, Oklahoma; or the price of Brent as traded on the Intercontinental Exchange (ICE, which the International Petroleum Exchange has been incorporated into) for delivery at Sullom Voe.

The price of a barrel (which is 42 gallons) of oil is highly dependent on both its grade (which is determined by factors such as its specific gravity or API and its sulphur content) and location. The vast majority of oil will not be traded on an exchange but on an over-the-counter basis, typically with reference to a marker crude oil grade that is typically quoted via pricing agencies such as Argus Media Ltd. and Platts.

For example in Europe a particular grade of oil, say Fulmar, might be sold at a price of "Brent plus US$0.25/barrel" or as an intra-company transaction. IPE claim that 65% of traded oil is priced off their Brent benchmarks. Other important benchmarks include Dubai, Tapis, and the OPEC basket. The Energy Information Administration (EIA) uses the Imported Refiner Acquisition Cost, the weighted average cost of all oil imported into the US as their "world oil price".

It is often claimed that OPEC sets the oil price and the true cost of a barrel of oil is around $2, which is equivalent to the cost of extraction of a barrel in the Middle East. These estimates of costs ignore the cost of finding and developing oil reserves. Furthermore the important cost as far as price is concerned, is not the price of the cheapest barrel but the cost of producing the marginal barrel.

By limiting production OPEC has caused more expensive areas of production such as the North Sea to be developed before the Middle East has been exhausted. OPEC's power is also often overstated. Investing in spare capacity is expensive and the low oil price environment in the late 90s led to cutbacks in investment. This has meant during the oil price rally seen between 2003-2005, OPEC's spare capacity has not been sufficient to stabilise prices.

Oil demand is highly dependent on global macroeconomic conditions, so this is also an important determinant of price. Some economists claim that high oil prices have a large negative impact on the global growth. This means that the relationship between the oil price and global growth is not particularly stable although a high oil price is often thought of as being a late cycle phenomenon.

A recent low point was reached in January 1999, after increased oil production from Iraq coincided with the Asian financial crisis, which reduced demand. The prices then rapidly increased, more than doubling by September 2000, then fell until the end of 2001 before steadily increasing, reaching

US $40 to US $50 per barrel by September 2004. In October 2004, light crude futures contracts on the NYMEX for November delivery exceeded US $53 per barrel and for December delivery exceeded US $55 per barrel.

Crude oil prices surged to a record high above $60 a barrel in June 2005, sustaining a rally built on strong demand for gasoline and diesel and on concerns about refiners' ability to keep up. This trend continued into early August 2005, as NYMEX crude oil futures contracts surged past the $65 mark as consumers kept up the demand for gasoline despite its high price. Crude oil futures peaked at a close of over $77 a barrel in July 2006, and in December 2006 at about $63. That is just about where they began the year 2006.

Individuals can now trade crude oil through online trading sites margin account or their banks through structured products indexed on the Commodities markets.

9

Religicn Science and the Environment

Religion, Science and the Environment is a non profit organization conceived and established by His All Holiness Bartholomew, Ecumenical Patriarch of the Christian Orthodox Church of Constantinople. The organisation's mission is to raise awareness of pressing environmental issues and create a wider involvement in environmental matters by mobilising the world's religious leaders, scientists and politicians.

The movement operates through Symposium study voyages that take place on ships and that involve around 150 people at a time. The aim of the symposia is tu raise awareness on environmental issues through extensive panel discussions.

Apart from discussing the various issues, the participants also visit sites of special concern, meet officials and NGO representatives in the countries visited, propose solutions and initiate plans or institutions for environmental cooperation and education. Past symposia have taken place in Patmos, the Baltic sea, the Adriatic sea, the Danube, the Black sea and the Amazon river.

The driving force of the RSE symposia is the belief that the analytical tools of science and the spiritual messages of religion must come together if the earth's environment is to be safeguarded against further destruction. Through the symposia RSE has managed to attract funding regarding the environment from international institutions, to initiate regional cooperation among governments and non-governmental organisations, to found the Halki Ecological Institute (an educational initiative to increase awareness and cooperation on environmental issues) and make possible environmental training for regional clergy, journalists, educators and youth.

EUROPEAN SCIENCE AND ENVIRONMENT FORUM

The European Science and Environment Forum (ESEF), now defunct, called itself "an independent, non-profit-making alliance of scientists whose aim is to ensure that scientific debates are properly aired, and that decisions which are taken, and action that is proposed, are founded on sound scientific principles." Typically this manifested itself in questioning the science upon which environmental safety regulations are based.

The Forum was linked, via shared staff (Julian Morris and Roger Bate) and a shared web server, to the International Policy Network and the Sustainable Development Network.

In 1996, Roger Bate approached R.J. Reynolds Tobacco Company for a grant of £50,000 to fund a book on risk, containing a chapter on passive smoking, but the grant request was denied and the money was never received. In 1997, the ESEF published What Risk? Science, Politics and Public Health, edited by Roger Bate which included a chapter on passive smoking. In 1998 the Academic Members of ESEF included Bruce N. Ames, Sallie Baliunas, Robert C. Balling, Jack Barrett, C.J.F. Böttcher, Peter Dietze, Tor Ragnar Gerholm, Gerhard Gerlich, Sherwood Idso, Helmut Metzner, Patrick J. Michaels, William Mitchell, Harry N.A. Priem, Michel Salomon, S. Fred Singer, Willie Soon, Wolfgang Thüne, and Gerd-Rainer Weber, while Richard S. Courtney and Michael Gough were Business Members.

WEB-BASED INQUIRY OF SCIENCE ENVIRONMENT (WISE)

The Web-based Inquiry Science Environment (WISE) provides a platform for creating inquiry science projects for middle school and high school students to work collaboratively using evidence and resources from the Web. WISE inquiry projects are typically developed by teams including teachers, researchers, and scientists. These projects range in duration from 2 days to 4 weeks, suitable for teachers in every science topic at all grade levels. The use of Internet materials provides the foundation of WISE. All projects make use of some content from the World Wide Web, as well as additional Web pages authored for purposes of the project. This approach helps students learn to use the Internet for inquiry, critiquing Web sites, designing approaches, or comparing arguments. Typical projects engage students in designing solutions to problems (e.g., building a desert house that is warm at night and cool during the day), debating contemporary science controversies (e.g., the causes of declining amphibian populations), investigating scientific phenomena (e.g., thermal equilibrium in the classroom), or critiquing scientific claims found in web sites (e.g., arguments for life on Mars.)

WISE projects can also incorporate Java applets to enable online discussions, data collection, drawing, argument creation, resource sharing, concept mapping and other built-in components. WISE is entirely browser-based, meaning that students only need access to a computer with an Internet connection, with no required software other than the Web browser). All student work is saved on central project servers that enable student accounts and teacher accounts to be coordinated, with special Web environments designed to support teachers and students. Students can access their work from any computer on the Internet. Teachers can choose from the library of curriculum projects in the WISE Teacher's Portal, each accompanied by a set of materials including a detailed lesson plan, pre and post assessments, connections to the AAAS National standards, tools for setting up a custom grading scheme, and even a software tool that enables customization of the WISE project for local issues, geographical features or student populations. Teachers can monitor and grade student work, provide formative feedback during a project run, and manage their student accounts.

Finally, WISE provides a user-friendly interface to facilitate the authoring of new projects, with numerous technology-based components including online discussions, data collection, drawing, argument creation, resource sharing, concept mapping and other built-in tools. Developers can also include custom tools of their own design. Projects are further customizable by teachers and other end users through the same interface to better meet the needs of their students and constraints of local educational settings. The WISE Web site provides rich descriptions of the WISE learning environment, a library of inquiry projects for students and teachers, a teacher support center, and the complete authoring environment.

SCIENCE, TECHNOLOGY, SOCIETY AND ENVIRONMENT EDUCATION

Science, Technology, Society and Environment (STSE) Education, originates from the Science Technology and Society (STS) movement in science education. This is an outlook on science education that emphasizes the teaching of scientific and technological developments in their cultural, economic, social ad political contexts. In this view of science education, students are encouraged to engage in issues pertaining to the impact of science on everyday life and make responsible decisions about how to address such issues (Solomon, 1993 and Aikenhead, 1994).

Historical Context

STS is rooted in the postmodernist view of science (Fensham 1985; Pedretti 1997 and Aikenhead, 2003). This sees science as a human endeavour, embedded in the social, economic and political contexts in which scientific developments occur, rather than as set of theories, facts or observable outcomes (Solomon, 1993 and Bingle & Gaskell, 1994). This article provides information about the evolution of STSE, along with the theoretical and pedagogical implications of including STSE perspectives in science education.

Science Technology and Society (STS)

The STS movement has a long history in science education reform, and embraces a wide range of theories about the interface between science, technology and society (Solomon and Aikenhead, 1994; Pedretti 1997). Over the last twenty years, the work of Peter Fensham, the noted Australian science educator, is considered to have heavily contributed to reforms in science education. Fensham's efforts included giving greater prominence to STS in the school science curriculum (Aikenhead,

2003). The key aim behind these efforts was to ensure the development of a broad-based science curriculum, embedded in the socio-political and cultural contexts in which it was formulated. From Fensham's point of view, this meant that students would engage with issues concerning the impact of science and technology on everyday life from different viewspoints. They would also understand the relevance of scientific discoveries, rather than just concentrate on learning scientific facts and theories that seemed distant from their realities (Fensham, 1985 & 1988).

However, although the wheels of change in science education had been set in motion during the late 1970s, it was not until the 1980s that STS perspectives began to gain a serious footing in science curricula, in largely Western contexts (Gaskell, 1982). This occurred at a time when issues such as, animal testing, environmental pollution and the growing impact of technological innovation on social infrastructure, were beginning to raise ethical, moral, economic and political dilemmas (Fensham, 1988 and Osborne, 2000). There were also concerns among communities of researchers, educators and governments pertaining to the general public's lack of understanding about the interface between science and society (Bodmer, 1985; Durant et. al. 1989 and Millar 1996). In addition, alarmed by the poor state of scientific literacy among school students, science educators began to grapple with the quandary of how to prepare students to be informed and active citizens, as well as the scientists, medics and engineers of the future (e.g. Osborne, 2000 and Aikenhead, 2003). Hence, STS advocates called for reforms in science education that would equip students to understand scientific developments in their cultural, economic, political and social contexts. This was considered important in making science accessible and meaningful to all students and, most significantly, engaging them in real world issues (Fensham, 1985; Solomon, 1993; Aikenhead, 1994 and Hodson 1998).

Goals of STS

The key goals of STS focus on:

- An interdisciplinary approach to science education, where there is a seamless integration of economic, ethical, social and political aspects of scientific and technological developments in the science curriculum.

- Engaging students in examining a variety of real world issues and grounding scientific knowledge in such realities. In today's world such issues might include the impact on society of: global warming, genetic engineering, animal testing, deforestation practices, nuclear testing and environmental legislations, such as the EU Waste Legislation or the Kyoto Protocol.
- Enabling students to formulate a critical understanding of the interface between science, society and technology.
- Developing students' capacities and confidence to make informed decisions and take responsible action, to address issues arising from the impact of science on their daily lives.

Scope and Emphasis

Over the last two decades, STS curricula have taken a variety of forms. These emphasize a particular aspect of STS according to the socio-political environment in which they are formulated, as well as the particular views of curriculum developers on STS education and what is considered valid knowledge in a science curriculum (Solomon & Aikenhead 1994 and Aikenhead, 2003). For example, in Canada and Israel, STS goals directed towards understanding environmental issues were given greater emphasis. Hence, the addition of "E" to STS, producing STSE and STES respectively. Whereas in Belgium, goals focusing on ethics were given greater prominence in STS education, and resulted in the publication of the journal: Science Technologies Ethique Societé, (Aikenhead, 2003). However, for the most part, STS curricula are bound by an overarching curriculum framework. This reflects the three curriculum content areas for STS education described by Hodson (1998):

Learning Science and Technology: acquiring and developing conceptual and theoretical knowledge in science and technology, and gaining a familiarity with a range of technologies.

Learning About Science and Technology: developing an understanding of the nature and methods of science and technology, an awareness of the complex interactions among science, technology, society and environment, and a sensitivity to the personal, social and ethical implications of particular technologies. Doing Science and Technology: engaging in and developing expertise in scientific inquiry and problem solving; developing confidence and competence in tackling a wide range of "real world" technological tasks.

STSE Education

As mentioned before, STSE is a form of STS education, but places greater emphasis on the environmental consequences of scientific and technological developments. In STSE curricula, scientific developments are explored from a variety of perspectives, including economic, environmental, ethical, moral, social and political (Kumar and Chubin, 2000 & Pedretti, 2005). Therefore, there is no one single definition for STSE education. At best, it can be loosely defined as a movement that attempts to bring about an understanding of the interface between science, society, technology and the environment. A key goal of STSE is to help students realize the significance of scientific developments in their daily lives and foster a voice of active citizenship (Pedretti & Forbes, 2000).

A Way of Improving Scientific Literacy

Over the last two decades, STSE education has taken a prominent position in the science curricula of different parts of the world, such as Australia, Europe, the UK and USA (Kumar & Chubin, 2000). In Canada, the inclusion of STSE perspectives in science education has largely come about as a consequence of the Common Framework of science learning outcomes, Pan Canadian Protocol for collaboration on School Curriculum (1997):. This document highlights a need to develop scientific literacy in conjunction with understanding the interrelationships between science, technology and environment. According to Osborne (2000) & Hodson (2003), scientific literacy can be perceived in four different ways:

- *Cultural:* Developing the capacity to read about and understand issues pertaining to science and technology in the media.

- *Utilitarian:* Having the knowledge, skills and attitudes that are essential for a career as scientist, engineer or technician.
- *Democratic:* Broadening knowledge and understanding of science to include the interface between science, technology and society.
- *Economic:* Formulating knowledge and skills that are essential to the economic growth and effective competition within the global market place.

Rationale and Goals

In the context of STSE education, the goals of teaching and learning are largely directed towards engendering cultural and democratic notions of scientific literacy. Here, advocates of STSE education argue that in order to broaden students understanding of science, and better prepare them for active and responsible citizenship in the future, the scope of science education needs to go beyond learning about scientific theories, facts and technical skills. Therefore, the fundamental aim of STSE education is to equip students to understand and situate scientific and technological developments in their cultural, environmental, economic, political and social contexts (Solomon & Aikenhead, 1994; Bingle & Gaskell, 1994; Pedretti 1997 & 2005). For example, rather than learning about the facts and theories of weather patterns, students can explore them in the context of issues such as global warming. They can also debate the environmental, social, economic and political consequences of relevant legislations, such as the Kyoto Protocol. This is thought to provide a richer, meaningful and relevant canvas against which scientific theories and phenomena relating to weather patterns can be explored (Pedretti et. al. 2005). In essence, STSE education aims to develop the following skills and perspectives (Aikenhead, 1994; Pedretti, 1996; Alsop & Hicks, 2001):

- Social responsibility
- Critical thinking and decision making skills
- The ability to formulate sound ethical and moral decisions about issues arising from the impact of science on our daily lives

Knowledge, skills and confidence, to express opinions and take responsible action to address real world issues in science.

Curriculum Content

Since STSE education has multiple facets, there are a variety of ways in which it can be approached in the classroom. This offers teachers a degree of flexibility, not only in terms of incorporating STSE perspectives into their science teaching, but integrating other curricular areas such as history, geography, social studies and language arts (Richardson & Blades, 2001). The different approaches to STSE education described in the literature (Ziman, 1994 & Pedretti, 2005):

A way of Humanizing Science

This approach examines the history of science through concrete examples, and is viewed as way of demonstrating the fallibility of science and scientists.Learning about inventions or scientific theories through the lives and worlds of famous scientist. Students can research their areas of interest and present them through various activities: e.g. drama-role play, debates or documentaries. Through this kind of exploration, students examine the values, beliefs and attitudes that influenced the work of scientists, their outlook on the world, and how their work has impacted our present circumstances and understanding of science today.

Philosophical: Helps students formulate an understanding of the different outlooks on the nature of science, and how differing viewpoints on the nature and validity of scientific knowledge influence the work of scientists. Thus, demonstrating how society directs and reacts to scientific innovation.

Using historical narratives or stories of scientific discoveries to concretely examine philosophical questions and views about science. For example, "The Double Helix" by James D. Watson is an account of the discovery of DNA. This historical narrative can be used to explore questions such as: "What is science? What kind of research was done to make this discovery? How did this scientific development influence our lives? Can science help us understand everything about our world?" Such an exploration embeds philosophical debates, pertaining to the nature of science, in their social and historical context. Thus making this kind of scientific inquiry concrete, meaningful and applicable to students' realities.

Issues-Based

This is the most widely applied approach to STSE education. It is directed towards formulating an understanding of the science behind issues along with the consequences to society and the environment. A multi-faceted approach to examining issues highlights the complexities of real-life debates. Students also become aware of the various motives influencing decisions made to address environmental issues.Real life events within the community, at the national or international level can be examined from political, economic, ethical and social perspectives through presentations, debates, role-play, documentaries and narratives. Real life events might include: the impact of environmental legislations, industrial accidents and the influence of particular scientific or technological innovations on society and the environment.

Opportunities and Challenges of STSE Education

Although advocates of STSE education keenly emphasize the merits of using this approach in science education, they also recognize the inherent difficulties that accompany its implementation. The opportunities and challenges of STSE education have been articulated by Hughes (2000) and Pedretti & Forbes, (2000), and are located at five different levels, as described below:

Values and Beliefs: STSE education, by nature of its goals, may challenge the values and beliefs of students, teachers, as well as conventional, culturally entrenched, views on scientific and technological developments. In many ways, this provides students with opportunities to engage with, and deeply examine the impact of scientific development on their lives from a critical and informed perspective. This not only helps to develop students' analytical and problem solving capacities, but also the ability to make informed choices in their everyday lives. Therefore, to enable students to formulate their own thoughts, independently explore other opinions and have the confidence to voice their personal viewpoints, teachers need to provide a balanced view of the issues being explored as they plan and implement STSE education lessons. They also need to cultivate safe, non-judgemental classroom environments, and must also be careful not to impose their own values and beliefs on students.

Knowledge and Understanding: The interdisciplinary nature of STSE education requires teachers to research and gather information from a variety of sources. At the same time, they need to develop a sound understanding of issues from various disciplines including: philosophy, history, geography, social studies, politics, economics, environment and science. This is so that students' knowledge base can be appropriately scaffolded to enable them to effectively engage in discussions, debates and decision-making processes. However, most science teachers are specialized in a particular field of science. In addition, lack of time and resources may effect how deeply teachers and students can examine issues from multiple perspectives. Nevertheless, a multi-disciplinary approach to science education enables students to gain a more rounded perspective on the dilemmas, as well as the opportunities, that science presents in our daily lives.

Pedagogic Approach: Depending on teacher's experience and comfort-level, a variety of pedagogic approaches, based on constructivism, can be used to make STSE education come alive in the classroom. The pedagogies used in STSE classrooms need to take students through different levels of understanding to develop their abilities and confidence to critically examine issues and take responsible action. Here, teachers are often faced with the challenge of transforming classroom pactices from task orientated approaches to those focussing on developing students' understanding and transferring agency for learning to students (Hughes, 2000). A compilation of pedagogic approaches for STSE education described in the literature (e.g. Hodson, 1998; Pedretti & Forbes 2000; Richardson & Blades, 2001):

1. Appreciating the societal impact of scientific and technological change and recognizing that, to some extent, science and technology are culturally determined. Students work in groups to research information on

various aspects of an event or innovation to illustrate its impact on society e.g. biotechnology, nuclear testing. Students can use role-play, concept mapping, posters, gallery exhibitions, presentations or documentaries to display research findings and express their own viewpoints.

2. Recognizing that decisions about scientific and technological development are taken in pursuit of particular interests. Weighing out the pros and cons of scientific and technological developments and their links with wealth and power. Debates or Town Hall style meetings, role-play interviews, case-studies, seminars, multimedia or documentary style presentations, can be used to present the political, social, scientific and economic factors that led to decisions about the impact of a particular scientific or technological development.
3. Developing one's views and establishing personal value positions on the effect of a scientific or technological development. Students can present their opinions on issues through: presentations, debates, group discussions, poster presentations and writing. Through such activities, students can also be encouraged to express their hopes, concerns and decisions as informed citizens.
4. Preparing for taking action. Having examined the complexity of the development, students explore and plan ways of addressing the issues. This is a fundamental aspect of STSE education and could involve: letter writing campaigns, writing a letter to the editor of a newspaper, developing a Webpage, presenting debates or holding meetings for the local community, developing personal action plans.

Time and Resources: The multi-faceted approach of STSE education requires teachers to move much beyond using conventional curriculum materials, and explore other resources in different curricular areas. This includes: social geography, history, social studies and politics. Collecting such resources, developing one's own background knowledge and integrating other curricular areas for successful and effective STSE lesson planning, require time and effort on behalf of the teacher.

Assessment and Evaluation: The broad, inquiry based approach to STSE education requires tools that assess students understanding of issues and the development of skills (e.g. problem solving, analysis, communication, presentation), rather than their decisions or opinions. Hence, STSE education calls for the use of qualitative rather than quantitative assessment methods. It is also difficult to develop assessment or evaluation criteria for such a personalized, objective, view of science. Here, teachers need to make students clearly aware that it is their efforts and skill development that are being assessed, rather than opinions. Examples of assessment tools might include: quizzes, questionnaires, journal writing, development of portfolios, observations and one-on-one exit interviews.

BROOKLYN ACADEMY FOR SCIENCE AND THE ENVIRONMENT (BASE)

BASE or Brooklyn Academy for Science and the Environment is a science high school. In 2003, the Prospect Park Alliance and the Brooklyn Botanic Garden partened up to establish BASE. BASE uses field study in Prospect Park and the Brooklyn Botanic Garden to supplement classroom teaching about science, nature, and the environment. BASE shares the Prospect Heights Campus with 4 other small schools. BASE, a unique school has 3 campuses. The other two are Prospect Park and the Brooklyn Botanic Garden.

NATIONAL COUNCIL FOR SCIENCE AND THE ENVIRONMENT

This article is uncategorized. Please help improve this article by adding it to one or more categories, so it may be associated with related articles (how?). Please remove this tag after categorizing, but not before. This page has been tagged since December 2006.

The National Council for Science and the Environment (NCSE) is a US-based non-profit organization which has a mission to improve the scientific basis for environmental decisionmaking. While an advocate for science and its use,

NCSE does not take positions on environmental policies. NCSE specializes in programs that bring together diverse individuals, institutions and communities to collaborate; particularly the many communities creating and using environmental knowledge, including research, education, environmental, and business organizations, as well as governmental bodies at all levels.

NCSE operates programs in five areas:

1. Science Solutions-programs that engage different communities to deploy science for decision-making on specific topics. This includes the National Commission on Science for Sustainable Forestry and the Wildlife Habitat Policy Research Program.
2. Strengthening education-which includes a University Affiliate Program (~ 140 member schools) with a Council of Environmental Deans and Directors and Environmentors, a high school mentoring program
3. National Conference on Science, Policy and the Environment
4. The Earth Portal-which includes the Encyclopedia of Earth. NCSE acts as the Secretariat for the Environmental Information Coalition which is building the Earth Portal. The Earth Portal is part of the Digital Universe project.
5. Science Policy-which aims to build understanding of, and support for, environmental science and its application, and for the programs that make it possible. NCSE played an instrumental role in stimulating an National Science Foundation initiative to triple its annual budget for environmental research, education, and scientific assessment.

10

Environmental Movement

The Environmental Movement (a term that sometimes includes the conservation and green movements) is a diverse scientific, social, and political movement. In general terms, environmentalists advocate the sustainable management of resources and stewardship of the natural environment through changes in public policy and individual behavior.

In its recognition of humanity as a participant in (not enemy of) ecosystems, the movement is centered around ecology, health, and human rights. The environmental movement is represented by a range of organizations, from the large to grassroots. Due to its large membership, varying and strong beliefs, and occasionally speculative nature, the Environmental Movement is not always united in its goals.

At its broadest, the movement includes private citizens, professionals, religious devotees, politicians, and extremists. Environmentalists are also often linked with other social movements, such as human and animal rights and pacifism.

REGIONAL ENVIRONMENTAL MOVEMENTS

- Environmental movement in the United States
- Environmental movement in New Zealand

History

The Environmental history of the United States: The Early Conservation Era is between (1832-1960). During the first years there were two early conservationist that made an impact, Henry David Thoreau and George Perkins Marsh. Thoreau was concerned about the wildlife from Massachusetts. He wrote Walden; or, Life in the Woods as he studied the wildlife from a cabin. George Perkins March was influential about the need for resource conversation.

The roots of the modern environmental movement can be traced to attempts in nineteenth-century Europe and North America to expose the costs of environmental negligence, notably disease, as well as widespread air and water pollution, but only after the Second World War did a wider awareness begin to emerge.

During the 1950s, 1960s, and 1970s, several events illustrated the magnitude of environmental damage caused by man. In 1954, the 23 man crew of the Japanese fishing vessel Lucky Dragon was exposed to radioactive fallout from a hydrogen bomb test at Bikini Atoll. In 1962 the publication of the book Silent Spring by Rachel Carson drew attention to the impact of chemicals on the natural environment. In 1967 the Torrey Canyon oil tanker went aground off the southwest coast of England, and in 1969 oil spilled from an offshore well in California's Santa Barbara Channel. In 1971 the conclusion of a law suit in Japan drew international attention to the effects of decades of mercury poisoning on the people of Minamata.

At the same time, emerging scientific research drew new attention to existing and hypothetical threats to the environment and humanity. Among them were Paul R. Ehrlich, whose book The Population Bomb, published 1968, revived concerns about the impact of exponential population growth. Biologist Barry Commoner generated a debate about growth, affluence and "flawed technology." Additionally, an association of scientists and political leaders known as the Club of Rome published their report The Limits to Growth in 1972, and drew attention to the growing pressure on natural resources from human activities. Meanwhile, nuclear proliferation and photos of Earth from space emphasized the consequences of technological accomplishments, as well as Earth's truly small place in the universe.

In 1972, the United Nations Conference on the Human Environment was held in Stockholm, and for the first time united the representatives of multiple governments in discussion relating to the state of the global environment. This conference led directly the creation of government environment agencies and the UN Environment Program. The United States also passed new legislation such as the Clean Water Act, the Clean Air Act, the Endangered Species Act, and the National Environmental Policy Act-the foundations for current environmental standards.

Love Canal

In 1978, a major environmental catastrophe at Love Canal, New York led to the Comprehensive Environmental Response, Compensation, and Liability Act (CERCLA) commonly called the "Superfund Act."

In the Love Canal incident, the local school board purchased land near Hooker Chemical's toxic burial site for $1 and constructed a new elementary school. In 1978, resident Lois Gibbs led an effort to investigate community concerns about the health of its residents. Neighborhood residents had observed strange odors and substances in the area, an extremely high rate of cancer, and an alarming number of birth defects. Gibbs' own children were hospitalized regularly for epilepsy, asthma, and urinary tract infections.

Because nobody was willing to pay for a solution to the problem, Ms. Gibbs struggled to gain attention until President Jimmy Carter announced a federal emergency. However, scarcity of government funds allowed for only residents nearest to the contaminated area be evacuated.

Later, when more money was secured, the government relocated and reimbursed more than 800 families. More than $200 million was used to clean up the site, and Congress passed the Superfund Act, which holds polluters accountable for misdeeds. So-called superfund sites are widespread in the United States, and are documented on a website run by the EPA.

Since the 1970s, public awareness, environmental sciences, ecology, and technology have advanced to include modern focus points like ozone depletion, global climate change, acid rain, and the harmful potential of GMOs.

SCOPE OF THE MOVEMENT

Biological Studies

- Environmental science is the study of the interactions among the physical, chemical and biological components of the environment;
- Ecology, or ecological science, is the scientific study of the distribution and abundance of living organisms and how these properties are affected by interactions between the organisms and their environment.

Primary Focus Points

- The Environmental Movement is broad in scope and can include any topic related to the environment, conservation, and biology, as well as preservation of landscapes, flora, and fauna for a variety of purposes and uses.
- The Conservation movement seeks to protect natural areas for sustainable consumption, as well as traditional (hunting, fishing, trapping) and spiritual use.

Other Focus Points

- Environmental health movement dates at least to Progressive Era, and focuses on urban standards like clean water, efficient sewage handling, and stable population growth. Environmental health could also deal with nutrition, preventive medicine, aging, and other concerns specific to human well-being. Environmental health is also seen as an indicator for the state of the natural environment, or an early warning system for what may happen to humans.
- Environmental Justice is a movement that began in the U.S. in the 1980s and seeks an end to environmental racism and prevent low-income and minority communities from an unbalanced exposure to highways, garbage dumps, and factories. The Environmental Justice movement seeks to link "social" and "ecological" environmental concerns, while at the same time preventing de facto racism, and classicism.

- Ecology movement could involve the Gaia Theory, as well as Value of Earth and other interactions between humans, science, and responsibility.
- Deep Ecology is often considered to be a spiritual spinoff of the ecology movement.
- Bright green environmentalism is a currently popular sub-movement, which emphasizes the idea that through technology, good design and more thoughtful use of energy and resources, people can live responsible, sustainable lives while enjoying prosperity.

ENVIRONMENTAL LAW AND THEORY

Property Rights

Many environmental lawsuits question the legal rights of property owners, and whether the general public has a right to intervene with detrimental practices occurring on someone else's land.

Citizens' Rights

One of the earliest lawsuits to establish that citizens may sue for environmental and aesthetic harms was Scenic Hudson Preservation Conference v. Federal Power Commission, decided in 1965 by the Second Circuit Court of Appeals. The case helped halt the construction of a power plant on Storm King Mountain in New York State.

"Nature's Rights"

Christopher D. Stone's 1972 essay, "Should trees have standing?" addressed the question of whether natural objects themselves should have legal rights. In the essay, Stone suggests that his argument is valid because many current rights-holders (women, children) were once seen as objects. However, his essay is still often regarded as logical fallacy.

ENVIRONMENTAL REACTIVISM

Numerous criticisms and ethical ambiguities have led to growing concerns about technology, including the use of potentially-harmful pesticides, water additives like fluoride, and the extremely dangerous ethanol-processing plants. NIMBY syndrome refers to public outcry caused by knee-jerk reaction to an unwillingness to be exposed to even necessary developments. Some serious biologists and ecologists created the scientific ecology movement which would not confuse empirical data with visions of a desirable future world.

Modern Environmentalism

Today, the sciences of ecology and environmental science, rather than any aesthetic goals, provide the basis of unity to most serious environmentalists. As more information is gathered in scientific fields, more scientific issues like biodiversity, as opposed to mere aesthetics, are a concern. Conservation biology is rapidly-developing field. Environmentalism now has proponents in business: New ventures such as those to reuse and recycle technical equipment are becoming more and more popular. Computer liquidators are just one example.

Criticisms of the Environmental Movement

Skeptics believe that environmentalism is more deeply rooted in politics than science, citing an uneasiness about compromising capitalistic ideals, or a sacrifice of a comfortable lifestyle. These concerns are often made valid by extremists associated with the movement who overshadow real accomplishments and scientific findings. One recent critic is the American author Michael Crichton, who appeared before the U.S. Senate Committee on Environment and Public Works to address concerns illustrated in his novel that questions the existence of the global climate crisis. Mr. Crichton recommended use of double-blind experimentation in environmental research, which is a standard in medicine and many other scientific fields. A consistent theme acknowledged by both supporters and critics is that scientists are constantly learning more about nature's complexities, and that the environmental debate deals with many points of view. Environmental studies are also criticized for the reliance on recent data to define long-term environmental trends. Because of this, many environmentalists support the precautionary principle, or to err on the side of caution when making public policy. This conjecture, however, is precisely what makes skeptics nervous.

Bibliography

Angel, Bradley: *Toxic Threat to Indian Lands: A Greenpeace Report*, San Francisco, Greenpeace, 1991.

Bailey, Conner and Charles E. Faupel: *Environmentalism and Civil Rights in Sumter County*, Colorado, Westview Press, 1992.

Bailey, Conner, Kelly Alley, Charles E. Faupel, and Cathy Solheim: *Environmental Justice: Issues, Policies, and Solutions*, Washington, D.C., Island Press, 1995.

Barnett, P.G.: *Survey of Research on the Impacts of Pesticides on Agricultural Workers and the Rural Environment*, Davis, CA: California Institute of Rural Studies, 1989.

Berry, Brian L.: *The Social Burdens of Environmental Pollution: A Comparative Metropolitan Data Source*, Cambridge, Ballinger Publishing Co., 1977.

Brown, Phil and Edwin J. Mikelson: *No Safe Place: Toxic Waste, Leukemia, and Community Action*, Berkeley, University of California Press, 1990.

Bryant, Bunyan and Paul Mohai: *Race and the Incidence of Environmental Hazards: A Time for Discourse*, Boulder, Westview Press, 1992.

Bryant, Bunyan: *Environmental Justice: Issues, Policies, and Solutions*, Washington, DC, Island Press, 1995.

Bullard, Robert D.: *Unequal Protection: Environmental Justice and Communities of Color*, San Francisco, Sierra Club Press, 1994.

Burch, William R. Jr.: *Human Ecology: An Environmental Approach*, Belmont, Duxbury Press, 1994.

Campbell, R. R., and Wade, J. L.: *Society and Environment: The Coming Collision*, Boston, Allyn and Bacon, 1972.

Churchill, Ward: *Struggle for the Land: Indigenous Resistance to Genocide, Ecocide, and Expropriation in Contemporary North America.* Monroe, Common Courage Press, 1993.

Coggin, Terrance P. and John M. Seidl: *Politics American Style: Race, Environment and Central Cities*, New York, Prentice-Hall, 1972.

Dasmann, R.F.: *Environmental Conservation,* New York, Wiley, 1972.

Daubenmire, R. F.: *Plants and Environment*, New York, Wiley, 1947.

Edelstein, Michael R.: *Contaminated Communities: The Social and Psychological Impacts of Residential Toxic Exposure*, Boulder, Westview Press, 1987.

Erickson, Kai: *A New Species of Trouble: Explorations in Disaster, Trauma, and Community*, New York, W.W. Norton, 1994.

Fitton, Laura: *A Study of the Correlation between the Siting of Hazardous Waste Facilities and Racial and Socioeconomic Characteristics*, Washington, DC, Center for Policy Alternatives, 1992.

Freeman, A. Myrick: *Environmental Quality Analysis: Theory and Method in the Social Sciences*, Baltimore, John Hopkins University Press, 1972.

Gay, Kathlyn: *Saving the Environment: Debating the Costs*, New York, Franklin Watts, 1996.

Gedicks, Al.: *The New Resource Wars: Native and Environmental Struggles Against Multinational Corporations*, Boston, South End Press, 1993.

Gelobter, Michael: *The Distribution of Air Pollution by Income and Race*, Berkeley, University of California, 1989.

Gerrard, Michael B.: *Whose Backyard, Whose Risk: Fear and Fairness in Toxic and Nuclear Waste Siting*, Boston, MIT Press, 1994.

Glickman, Theodore S. and Robert Hersh: *Evaluating Environmental Equity: The Impacts of Industrial Hazards on Selected Social Groups in Allegheny County, Pennsylvania.* Washington, DC, Resources for the Future, 1995.

Goldman Benjamin A.: *Not Just Prosperity: Achieving Sustainability With Environmental Justice*, Washington, Corporate Conservation Council, 1993.

Gould, J. M.: *Quality of Life in American Neighborhoods: Levels of Affluence, Toxic Waste, and Cancer Mortality in Residential Zip Code Areas*. Boulder, CO, Westview Press, 1986.

Harrison, D.: *Who Pays for Clean Air: The Cost and Benefit Distribution of Automobile Emission Standards*, Cambridge, Ballinger, 1975.

Hill, David: *The Quality of Life in America; Pollution, Poverty, Power, and Fear*, New York, Holt, Rinehart and Winston, 1973.

Hofrichter, Richard: *The Theory and Practice of Environmental Justice*, Philadelphia, New Society Publishers, 1993.

Hurley, Andrew: *Environmental Inequalities: Class, Race, and Industrial Pollution in Gary, Indiana, 1945-1980*. Chapel Hill, University of North Carolina Press, 1995.

Jaimes, Annette: *The State of Native America: Genocide, Colonization, and Resistance*, Boston, South End Press, 1992.

Johnson, Rebecca L.: *The Greenhouse Effect: Life on a Warmer Planet*, Minn., Lerner Publications, 1993.

Karliner, Joshua: *The Wmx Corporation, Hazardous Waste, and Global Strategies For Environmental Justice*, San Francisco, Political Ecology Group, 1994.

Kruvant, W.J.: *People, Energy, and Pollution*, Cambridge, Ballinger Publishing, 1975.

LaBalme, J.: *A Road to Walk, A Struggle for Environmental Justice*, Durham, Regulator Press, 1987.

Logan John R. and Harvey L. Molotoch: *Urban Fortunes: The Political Economy of Place*, Berkeley, University of California Press, 1987.

Mandelker, D.R.: *Environment and Equity: A Regulatory Challenge*, New York, McGraw Hill, 1981.

Merchant, Carolyn: *Radical Ecology: The Search for a Livable World*, New York, Routledge, 1992.

Meyer, Art: *Earth Keepers: Environmental Perspectives on Hunger, Poverty, and Injustice*, Scottsdale, Herald Press, 1991.

Middleton, Nick: *Atlas of Environmental Issues*, New York, Facts on File, 1989.

Ostheimer, John M. and Leonard G. Ritt: *Environment, Energy, and Black Americans*, Beverly Hills and London, Sage Publications, 1976.

Sachs, Aaron: *Eco-Justice: Linking Human Rights and the Environment*, Washington, World Watch Institute, 1995.

Schwab, Jim: *Deeper Shades of Green: The Rise of Blue-Collar and Minority Environmentalism in America*, San Francisco, Sierra Club Press, 1994.

Sontheimer, Sally: *Women and the Environment: Crisis and Development in the Third World*, New York, Monthly Review Press, 1991.

Strahler, A. N., and Strahler, A. H.: *Introduction to Environmental Science: Santa Barbara*, New York, Hamilton Publishing Company; 1974.

Szasz, Andrew: *Ecopopulism: Toxic Waste and the Movement for Environmental Justice*, Minneapolis, University of Minnesota Press, 1994.

Vallette, J.: *The International Trade in Wastes: A Greenpeace Inventory*, Washington, Greenpeace, 1989.

Wenz, Peter S.: *Environmental Justice*, Albany, State University of New York Press, 1988.

Zupan, J.M.: *The Distribution of Air Quality in the New York Region*, Baltimore, Johns Hopkins University Press, 1973.

Index

L

M

N

O

P

Q

R

□□□